AF289870

ANEURAL ORGANISMS
IN NEUROBIOLOGY

ADVANCES IN BEHAVIORAL BIOLOGY

Recent Volumes in this Series

ANEURAL ORGANISMS IN NEUROBIOLOGY

Edited by
Edward M. Eisenstein

Department of Biophysics
Michigan State University
East Lansing, Michigan

PLENUM PRESS · NEW YORK AND LONDON

Library of Congress Cataloging in Publication Data

Main entry under title:

Aneural organisms in neurobiology.

(Advances in behavioral biology; v. 13)
Based in part on the proceedings of a symposium held at the Winter Conference on Brain Research in Vail, Colo. in January 1972.
Includes bibliographies and index.
1. Neurobiology—Congresses. 2. Micro-organisms—Physiology—Congresses. I. Eisenstein, Edward M., ed. [DNLM: 1. Neurophysiology—Congresses. 2. Protozoa—Congresses. W3 AD215 v. 13 1972 / QX50 A579 1972]
QP351.A65   576'.11'8   74-28345
ISBN-13:978-1-4613-4475-9   e-ISBN-13:978-1-4613-4473-5
DOI: 10.1007/978-1-4613-4473-5

Based in part on the proceedings of the symposium
"Aneural Organisms: Their Significance for Neurobiology"
held at the Winter Conference on Brain Research
in Vail, Colorado, in January, 1972

© 1975 Plenum Press, New York
Softcover reprint of the hardcover 1st edition 1975

A Division of Plenum Publishing Corporation
227 West 17th Street, New York, N.Y. 10011

United Kingdom edition published by Plenum Press, London
A Division of Plenum Publishing Company, Ltd.
4a Lower John Street, London, W1R 3PD, England

Preface

This volume had its inception at a symposium entitled
<u>Aneural Organisms: Their Significance for Neurobiology</u> held at the
Winter Conference on Brain Research in Vail, Colorado, in January,
1972. The original participants were Drs. Epstein, Hamilton, Kung,
Wood and myself. However, since that time several other authors
(Drs. Applewhite, Chen, Diehn and Ettienne) were asked to contribute papers and all were asked to update their presentations so
as to present a broad perspective as to the role and significance
of aneural systems for investigating neurobiological problems.
This volume is the result of that effort.

I wish to thank Dr. Claude F. Baxter, the program chairman
of the Winter Conference, and his staff for their help, as well
as the contributors for their efforts. Great appreciation is due
Mrs. Sharon Loomis for her excellent work in preparing the manuscript
for publication.

It is our hope that this volume will demonstrate the usefulness
and advantages in exploiting aneural systems for the insights they
may yield in answering some of the fundamental neurobiological
questions facing us.

E.M. Eisenstein

Contents

ANEURAL SYSTEMS AND NEUROBIOLOGY:

A POINT OF VIEW

E.M. Eisenstein
Department of Biophysics
Michigan State University
East Lansing, Michigan 48823

It has become increasingly apparent in the last several years
that invertebrate neural systems have much to offer in the attempt
to understand complex mammalian systems (Wiersma, 1967). Two or
three generations ago this might have been a trivial statement as
neuroscientists of that bygone era often worked across the phylogenetic kingdom and were well aware of evolutionary continuity.
Somehow in the twentieth century an intellectual cleavage between
comparative and strictly mammalian neurobiology developed. Since
this gap has recently begun to close, perhaps it is not unreasonable at this time to ask whether even an aneural system might not
have important contributions to make in understanding more complex
neural systems.

Many of the phenomena we study in neurobiology are not unique
to nervous tissue but also occur in aneural systems. For example,
conduction and sensory transduction can be seen in protozoa. There
are striking similarities across phylogeny of many metabolic pathways as well as membrane structure and function. It is therefore
reasonable to expect some similarity in the mechanisms underlying
related phenomena in aneural and neural systems.

If one views the functions carried out by nervous systems as
having both intracellular as well as intercellular components then
one might view such work on aneural single cells as a strategy to
focus on the intracellular components uncomplicated by intercellular
considerations. The advantage of this approach, I feel, over that
of examining a single neuron in a complex neural matrix, is that
all the events of importance related to the phenomenon under study--
be it sensory transduction, control of motor activity or behavior

modification--occur in that one cell. A major stumbling block in
approaching such questions in neural systems, i.e., the role of any
one cell in a complex matrix of cells, is bypassed. However,
this simplification is not achieved without a price, since the
functions which occur in different cells in higher organisms now
all occur in one cell. All problems in neurobiology cannot be
solved using aneural systems but this approach may be the most
useful in many cases. The choice of a model system for study should
be a matter of strategy rather than chance.

I believe that some of the most general and fundamental processes of living systems--transduction of an environmental event
into a biological one, genetic control of behavior, coupling of
stimulus input and behavior output as well as questions related
to behavior modification--may be approachable by molecular studies
at this phylogenetic level in a way that they are not in more
complex neural systems. For instance, the problem of causally
relating specific cellular changes in the nervous system (e.g.,
membrane alterations) to the behavior of the organism has been an
extremely difficult one. An important thread running throughout
this book is that in aneural systems, particularly single-celled
ones, it is possible to examine directly questions about "membrane-
behavior" coupling. At this level the conceptual gap between
"molecular" and "behavioral" data is considerably reduced.

The purpose of this volume is to report to neurobiologists what
progress has been made at the aneural level and what the prospects
are for making advances into some of the fundamental problems we
normally investigate in nervous systems.

The authors have tried to be selective rather than exhaustive
in their discussions of work in their respective areas. They have
considered advantages and disadvantages of examining in aneural
model systems some of the general problems neurobiologists study.
Keeping in mind the fact that this book is being directed to neurobiologists who by and large do not routinely accept aneural systems
as being important systems for the problems they investigate, each
author has made some effort to conceptually bridge the gap between
these aneural systems and the nervous system. I think all of the
contributors, one way or another, accept as valid the idea that
their work bears importantly on neural phenomena.

The order of the papers in this volume is based on the assumption that aneural behavior, as in neurally mediated behavior, can
be dissected into sensory, integrative and motor components. While
many of the authors have touched on all three, their chapters have
been arranged somewhat arbitrarily according to emphasis. We
start with sensory processes (Wood), followed by those which have a
bearing on integrative processes between sensory input and motor

output (Kung and Diehn). The chapters on motor output emphasize ciliary control (Epstein and Chen) and contraction (Ettienne). Behavior modification follows next (Hamilton), with a concluding chapter on a broad comparison of plant and animal behavior (Applewhite).

ACKNOWLEDGEMENTS

The work in my laboratory in this area has been supported by NSF grant GB 23371 and MH grant 1 RO3 MH 18570-01 to E.M.E. and NIH general research support grant 5-501-RR-05656-04 to the College of Human Medicine, Michigan State University.

REFERENCE

Wiersma, C.A.G. (Ed.) 1967. <u>Invertebrate</u> <u>Nervous</u> <u>Systems</u>. <u>Their</u> <u>significance</u> <u>for</u> <u>mammalian</u> <u>neurophysiology</u>. The University of Chicago Press: Chicago and London.

PROTOZOA AS MODELS OF

STIMULUS TRANSDUCTION

David C. Wood
Department of Psychology
University of Pittsburgh
Pittsburgh, Pennsylvania 15260

CONSIDERATIONS ON THE USE OF PROTOZOA AS
MODELS OF STIMULUS TRANSDUCTION

To the casual observer one of the most interesting behavioral characteristics of protozoa is their sensitivity to stimuli. At the beginning of this century Jennings (1906) and Mast (1906) described in detail some responses of protozoa to stimuli: the "avoiding reaction" of swimming ciliates when they bump into objects, the swimming behavior which causes flagellates to collect in lighted areas and the responses of amoeba to mechanical stimuli and food particles. These observations leave no doubt that protozoa can sense events in their external environment and therefore must contain receptor mechanisms.

Evolutionary considerations coupled with certain experimental results provide an _a priori_ rationale for suggesting that these protozoan receptor mechanisms may be similar to metazoan receptor mechanisms. The uniformity of genetic and metabolic mechanisms in animals from all branches of the phylogenetic tree suggests commonalities will be discovered in other molecular mechanisms. The similarity of the ionic and membrane mechanisms producing action potentials in the squid axon (Hodgkin and Huxley, 1952) and the frog node of Ranvier (Frankenhaeuser, 1960) demonstrates that phylogenetic uniformities are also to be found in neurophysiological mechanisms. In the field of sensory physiology itself the ommatidia of _Limulus_ and the stretch receptors of crayfish have been exploited as model receptors and the results obtained from these preparations have been used to suggest physiological processes occuring in mammalian receptors. The subjects of these studies, mollusks, arthropods and vertebrates, represent very

divergent phylogenetic branches with few common ancestors. These
ancestors were probably rather simple organisms which were either
unicellular or were metazoans with few differentiated tissues. Presumably these common ancestors had the genetic information required
for stimulus reception and action potential production. Since
protozoa must have evolved from these same or closely related primordial organisms it seems likely that they too may possess many
membrane and molecular mechanisms which are of neurophysiological
interest. Of course this presumed similarity cannot be uncritically
accepted until sufficient data are assembled to allow a more
adequate comparison between protozoan and metazoan receptors.

On the basis of evolutionary considerations it was suggested
above that protozoa might serve as appropriate subjects for neurobiological research. The use of protozoa as models of neurobiological
processes was advanced with the expectation that the study of protozoa would provide certain experimental advantages. Many of these
advantages have been discussed elsewhere in this volume and, hope-
fully, others will become apparent during the course of this review.
It is often assumed that one of these advantages is the simplicity
of protozoa. However I would consider most protozoa, and contractile ciliates in particular, to be more complex than metazoan
neurons and receptor cells. This complexity derives from the fact
that individual protozoa can respond to a variety of stimuli and can
produce more than one response. Unlike metazoan neurons, protozoa
do not differentiate markedly during development and hence do not
become specialized through the loss of certain functional capacities.

To circumvent the problems produced by the multiplicity of protozoan receptor and effector mechanisms it seems most logical to
attempt to study a given protozoan receptor or effector mechanism
in isolation. Experimentally this means that one must measure
electrophysiological events produced by stimulation or directly
elicit motor responses by passing electric currents. If only
behavioral indices of stimulus reception are used then the multiplicity of events internal to the protozoan may prevent a clear
assignment of effects to given physiological processes. On the
other hand a clear separation of physiological processes can be
made through the use of electrophysiological techniques. As a
consequence of electrophysiological analysis it is often possible
to perfectly correlate a particular protozoan behavior with an electrophysiological event occurring within the organism. Once this
correlation has been made, the behavior can be used as an index of
the electrophysiological event and inferences about the effect of
stimuli, drugs, etc. on electrophysiological processes can be made
on the basis of behavioral data. However, in all cases these inferences should be supported by electrophysiological recordings.
As a consequence of these considerations, stimulus reception in
protozoa will be considered from an electrophysiological viewpoint.

From this point of view the primary function of a receptor cell
is the transduction of mechanical, photic or other stimulus events
into graded transmembrane receptor potentials. These transduction
processes will be considered according to the mode of stimulation
producing the behavioral and electrophysiological responses.

TRANSDUCTION OF MECHANICAL STIMULI
BY PROTOZOA

At least three protozoa, _Noctiluca,_ _Paramecium_ and _Stentor_
generate transmembrane graded potentials in response to mechanical
stimuli. In these three cases the graded receptor potentials are
also capable of eliciting action potentials or regenerative
potentials which trigger subsequent locomotor or bioluminescent
responses. The earliest studied of these protozoa is the marine
dinoflagellate _Noctiluca_ (Eckert, 1965 a,b). This protozoan
responds to a mechanical stimulus with a graded receptor potential
which, if sufficiently large, can trigger a propagated action
potential. In turn the action potential triggers the animal's
bioluminescent flash. _Paramecia_ also produce a graded receptor
potential in response to mechanical stimulation (Naitoh and Eckert,
1969; Eckert and Naitoh, 1972; Eckert, 1972). A graded regenerative
transmembrane potential is elicited by the receptor potential if it
exceeds a certain level of depolarization (Eckert, Naitoh and
Friedman, 1972). In turn this regenerative response elicits the
behavioral response of ciliary reversal. The final example is that
of the contractile protozoan _Stentor._ This ciliate also responds
to mechanical stimulation with a graded receptor potential which
may elicit an action potential leading to a rapid contraction
(Wood, 1970b). In view of these examples it is clear that behavioral
responses of these protozoa to incident stimuli are attributable to
the transduction of mechanical stimuli into a sequence of transmembrane potentials which activate the effector system. Superficially at least, these electrical events appear analogous to
potentials observed in metazoan mechanoreceptors.

All reports of protozoan mechanoreceptor potentials indicate
that their amplitude increases as the stimulus intensity increases
but no quantitative relations between these variables have been
reported. In attempting to determine the relation between receptor
potential amplitude and stimulus intensity a number of problems
beset the experimenter. Protozoa are not rigid bodies nor are they
embedded in a rigid tissue matrix; hence the orientation between the
animal and the stimulus probe is always subject to change and the
constancy of the stimulus is always in doubt. This undesirable
situation is considerably aggravated by the fact that protozoa are
motile and can change their orientation to the stimulus probe even
in the absence of applied stimuli.

Recently I have found it is possible to circumvent some of
these problems by using the recording electrode as a stimulus probe.
To make the electrode function in this dual manner, it is fastened
to a linear velocity transducer in series with a small adjustable
solenoid. By actuating the solenoid the transducer core and electrode tip can be moved by 0-50μ. The animal impaled on the electrode
is mechanically stimulated as the electrode is moved. In this
manner the constancy of the site of stimulation can be confirmed by
the constancy of the resistive-capacitive properties of the recording site and the stability of the responses produced. Surprisingly,
if microelectrodes of less than 30 MΩ are used, this stimulation
procedure produces no stimulus artifacts greater than 0.5mv. Only
such minimal stimulus artifacts are observed when the microelectrode
is run all the way through the experimental animal and the mechanical
stimulus applied. When stimulated in this manner, _Stentor_ produce
stable electrical and behavioral responses during stimulation periods
of 15 minutes and longer, if the interstimulus interval is kept
sufficiently long. The validity of this method of stimulation is
confirmed by the similarity in form and amplitude of receptor potentials produced by this technique to those elicited by a separate
mechanical probe.

The study of stimulus intensity-response functions is hindered
by the fact that the peak of the protozoan receptor potential may
be occluded by an action potential which is triggered before the
receptor potential reaches its maximum amplitude. Neither tetrodotoxin nor procaine block these protozoan action potentials as
they do in metazoan neurons. In _Stentor_, action potentials can
be blocked by the introduction of K^+ into the bathing medium;
however, this procedure also results in a diminution of the amplitude
of the receptor potentials and an increase in their rise time.
Despite this problem, introduction of K^+ into the bathing medium
has been used as a technique to generate stimulus intensity-response
functions.

Sample records of receptor potentials and a stimulus intensity-
response graph are presented below. As recorded from intravacuolar
electrodes in _Stentor_, mechanoreceptor potentials are seen as brief
(< 50 msec.) positive-going responses (Figure 1A). Their amplitudes
and rates of rise clearly increase as the velocity of the electrode
movement is increased. The relation between peek stimulus velocity
and receptor potential amplitude follows a negatively accelerated
function (Figure 1B). At present we have too little data to ade-
quately characterize these functions mathematically, though the
sample data presented in Figure 1B can be closely approximated by
a logarithmic function. Mechanoreceptor potentials in _Paramecia_
are also reported to saturate at high stimulus intensities (Naitoh
and Eckert, 1969; 1973). Therefore the relationship between these

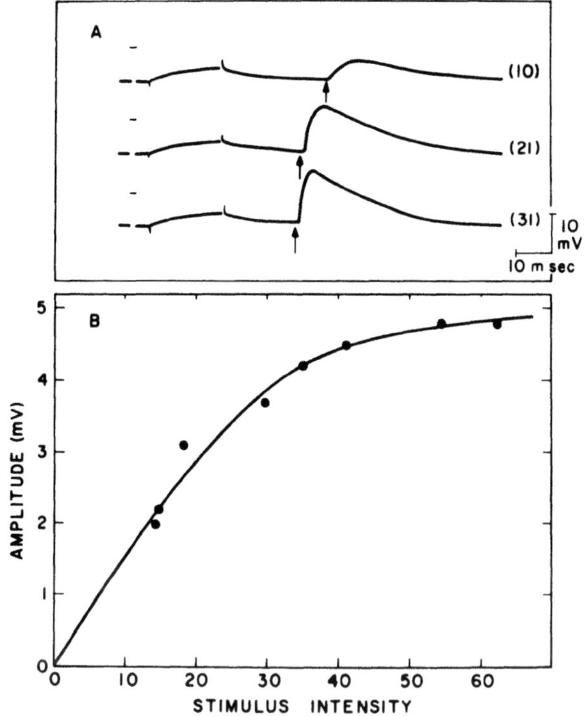

Figure 1 A. Three successive traces recorded from an intravacuolar
electrode in an animal being stimulated with varying stimulus intensities. Mechanical stimulus onset occurs at the arrow in each
case. The number to the right is the maximum output voltage in 0.1
mv units of a linear velocity transducer used to monitor the
stimulus velocity. The rectangular pulse on each trace is used to
measure electrode resistance. The second deflection is produced
by 1×10^{-9} amp. current passed through a Wheatstone bridge circuit
and is used to measure changes in membrane resistance. The mechanical stimulus appears earlier on records with higher stimulus intensities because the solenoid is actuated earlier by larger currents.
In all cases the intravacuolar "resting" potential is +2mv.

B. Sample graph of receptor potential amplitude against stimulus
intensity. Stimulus intensity is taken as the output of the linear
velocity transducer in mv. This data is from a different animal
than the records shown above.

receptor potentials and stimulus intensity must follow some form of
negatively accelerated curve similar to that described above in
Stentor.

Like metazoan mechanoreceptor potentials, protozoan receptor potentials appear to be generated in localized areas of the cell membrane which are specialized for stimulus reception. This membrane specialization is often evident in the animal's behavior, since mechanical stimulation applied to different areas of membrane may produce different responses. For example, mechanical stimulation of the anterior end of _Paramecium_ results in ciliary reversal which is produced by a depolarizing receptor potential which secondarily elicits a regenerative Ca^{++} spike (Figure 2A, B) (Eckert, Naitoh and Friedman, 1972). On the other hand stimulation of the posterior end of this animal results in an increased frequency of ciliary beating and faster forward swimming (Figure 2A). This latter behavioral response is activated by a hyperpolarizing receptor potential (Naitoh and Eckert, 1969). Thus the local nature of the mechanisms generating receptor potentials in _Paramecia_ is evident in both the different behaviors and different polarities of the receptor potentials produced by stimulation of opposite ends of the animal.

In _Stentor_, stimulation of the anterior portion of the animal is obviously more effective in eliciting contractions than is stimulation of the posterior portion. Even in the anterior portion of the animal, receptor potentials are larger when recorded from microelectrodes near the site of microprobe stimulation than when recorded from more distal electrodes (Wood, 1972). Action potentials are also initiated near the site of stimulation and propagated away from it. Thus, like _Paramecium_, _Stentor_ appears to generate mechanoreceptor potentials locally near the site of stimulation and to have membrane surfaces showing some degree of specialization for mechanoreception.

A number of lines of evidence indicate that protozoan receptor potentials are generated as a result of changes in membrane conductance for specific ions. The most direct line of evidence has been supplied by Naitoh and Eckert (1973) who were able to measure membrane resistances in _Paramecium_ with rectangular current pulses during the course of the posterior hyperpolarizing receptor potential. As expected the measured membrane resistance fell markedly after the onset of the receptor potential and increased to resting values as the receptor potential decayed. This same technique has been employed during the anterior depolarizing receptor potential with similar results (Eckert, Naitoh and Friedman, 1972). However regenerative Ca^{++} spikes are also elicited by anterior stimulation; consequently some or all of the decrease in membrane resistance may be related to the Ca^{++} spike production and not to the receptor potential.

Injected d.c. currents have been used to depolarize or hyperpolarize the plasma membrane of _Paramecium_ and _Stentor_. In _Paramecium_, depolarization of the plasma membrane progressively reduces the

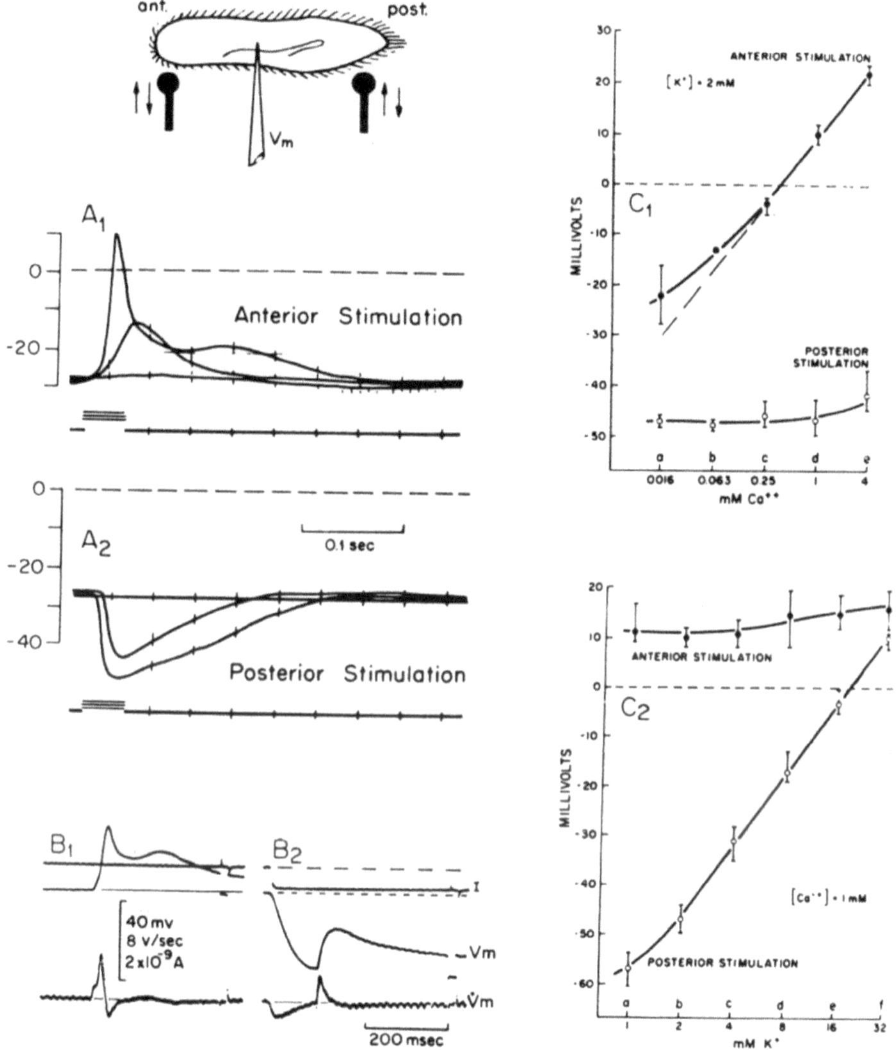

Figure 2. Electrical responses to mechanical stimulation in _Paramecium caudatum_. Stimuli applied with electrically driven microstylus to anterior (A_1) and posterior (A_2) surfaces. Lower traces in A_1 and A_2 show pulses energizing the microstylus. (B_1) Response to mechanical stimulation of cell anterior; (B_2) mechanical stimulus applied to same cell during hyperpolarizing current. The regenerative calcium response was thereby suppressed, leaving unobscured anterior receptor potential (I, current trace; V_m, membrane potential; $\dot{V}m$, first time derivation of V_m). Curves in C_1 show effects of extracellular calcium concentrations on peak values of potentials evoked by anterior and posterior mechanical stimulation. Curves in C_2 show effects of extracellular potassium concentrations. (Taken from Eckert, 1972).

amplitude and rate of rise of the anterior depolarizing receptor
potential, though this effect may be partially attributable to
increases in membrane conductance produced by the depolarization
(Eckert, Naitoh and Friedman, 1972). Conversely, hyperpolarization
of the membrane increases the rate of rise and amplitude of the
anterior receptor potential. In Stentor, the rate of rise of the
receptor potential is progressively decreased with increasing polarization of the membrane and with sufficiently large polarizations
the polarity of the receptor potential is reversed (Wood, 1973).
These results can be explained by assuming that the ions producing
these protozoan receptor potentials have a transmembrane equilibrium
potential which can be approached or reached by injecting d.c.
current thus minimizing or eliminating the ionic fluxes which produce
the receptor potentials. These results are compatible with models
used to explain the production of graded synaptic potentials in
neurons (Ginsborg, 1967).

Finally, the peak of the posterior receptor potential in
Paramecium varies as a function of extracellular K^+ in a manner
approximating the predictions of the Nernst equation (Figure 2C).
Since in the derivation of the Nernst equation it is assumed that
the membrane under study is perfectly permeable to ions which obey
this relationship, it appears that the posterior membranes of
Paramecium must become highly permeable to K^+ during the peak of the
posterior receptor potential. Anterior receptor potentials in
Paramecium and Stentor cannot be studied in this manner because
spike potentials occlude their peak amplitude.

As suggested above, protozoan receptor potentials appear to be
generated by transmembrane fluxes of specific ions. The peak of
the posterior receptor potential of Paramecium closely follows the
Nernst relation when extracellular K^+ concentrations are varied but
is not markedly altered by changing the extracellular concentration
of Ca^{++}. In Stentor the rate of rise of the receptor potential is
also a function of extracellular K^+ but is not altered by changes
in extracellular Mg^{++}, Na^+ or Cl^- (Wood, 1974). The anterior
receptor potential of Paramecium on the other hand is most dependent
on extracellular Ca^{++} concentration and not markedly affected by
extracellular K^+ (Figure 2C). Thus those protozoan receptor potentials which have been studied by changing the ionic milieu show
ionic selectivity.

Many of the particulars of mechanoreceptor potential production
cited above seem familiar because they are also characteristic of
metazoan receptor potential production. In both systems a localized
increase in membrane conductance for specific ions generates the
observed receptor potential. While metazoan receptor cells usually
show stimulus evoked increases in Na^+ conductance (along with Ca^{++}
and K^+ conductances in some cases) the fresh water ciliates which
have been investigated do not rely on an increase in sodium conductance

to produce receptor potentials. This may be related to the fact
that the protozoa studied live in fresh water where, unlike metazoan extracellular fluid, the concentration of Na^+ is too low to
allow a large Na^+ concentration gradient between the inside and
outside of the cell.

Despite the numerous similarities between protozoan and metazoan receptor potentials, continued study of protozoan receptor
systems is advantageous only if there are particular features of
protozoan receptor systems which make them particularly amenable
to study. The most obvious of these advantages is the accessibility
of protozoa to study. Coupled with the ease of obtaining isolated
protozoa is the ease of inferring electrophysiological events from
behavioral observations alone. For instance, contractions in
<u>Stentor</u> are always preceded by action potentials so the observation
of a contraction in this organism can be used to infer the presence
of an action potential. Agents affecting receptor potentials can
also be inferred on the basis of behavioral tests. Mechanical
stimuli elicit receptor potentials, action potentials and contractions in that order, while electrical stimuli elicit action potentials and contractions and bypass the receptor potential mechanism
(Wood, 1970b). Thus agents which alter <u>Stentor's</u> sensitivity to
mechanical stimuli but do not affect its electrical threshold must
act on the mechanism which generates the mechanoreceptor potential.

We have recently completed a series of drug studies employing
the above technique to study mechanoreceptor potential generation.
Animals were tested initially in varying concentrations of the drug
under study to determine the maximal concentration of drug which
produced no alterations in electrical threshold. This concentration
was then used in a study of mechanical stimulus sensitivity. In this
study animals were presented with a series of constant amplitude
stimuli to determine their probability of contracting to these
stimuli.

As can be seen in Table 1, 7 of the 34 agents tested did affect
mechanical stimulus sensitivity but not electrical thresholds. It
is striking that all the nicotinic cholinergic blocking agents
tested markedly depressed the probability of response to mechanical
stimuli. While dose-response curves have not yet been generated for
all these drugs, the dose-response curve for d-tubocurarine has
been found to follow a sigmoid curve between the concentrations of
0.001 and 0.03 mM. Conversely the cholinomimetic nicotine poten-
tiates the probability of response. This potentiation becomes
more obvious with repeated stimulation to which drugged animals
habituate less than controls (Wood, 1970a). After 15 minutes of
1/minute stimulation, animals in a medium containing 0.1mM nicotine
bitartrate had a response probability of 0.25 above that of the
controls. Two other agents, cocaine and atropine, produce slight
depressions in response probability.

TABLE 1

Drug	Concentration	Probability of Response
Control (15 groups)		0.93 – 0.50
General Stimulants		
Caffeine	0.3mM	0.84
Strychnine	0.01	0.62
Picrotoxin	0.3	0.87
General Sedatives		
Phenobarbital	0.1	0.87
Local Anesthetics		
Cocaine	0.3	0.39*
Procaine	0.3	0.68
Cholinergic Agents		
Nicotinic blocking agents		
Decamethonium Br	1.0	0.01*
d-tubocurarine	0.03	0.03*
Succinylcholine Cl	1.0	0.13*
Tetra ethylammonium Cl	1.0	0.21*
Muscarinic blocking agents		
Atropine sulfate	0.03	0.44*
Scopolamine	1.0	0.69
Mimetics		
Acetylcholine Cl	1.0	0.52
Carbachol	1.0	0.62
Nicotine	0.1	0.97*
Pilocarpine	1.0	0.81
Anticholinesterase agents		
Eserine	0.1	0.56
Neostigmine	1.0	0.81
Adrenergic Agents		
Blocking agents		
Phenoxybenzamine	0.01	0.69
Pindolol	0.3	0.75
Mimetics		
Amphetamine	0.03	0.80
Isoproterenol	0.03	0.64
Norepinephrine	0.1	0.50

Drug	Concentration	Probability of Response
Putative Transmitters		
Dopamine	0.03	0.60
Gaba	1.0	0.74
Glutamic acid	1.0	0.76
Glycine	1.0	0.70
Miscellaneous Agents		
Digitoxin	1.0	0.75
Ouabain	1.0	0.93
Pyrilamine	0.003	0.63
Hemicholinium	1.0	0.79

* Groups whose probability of response falls outside the range of
the controls. The highest concentration of drug was used which did
not increase the animals threshold for electrical shock. All
groups involve 25 or more animals.

Applewhite (1972) has tested the effect of 0.05mM d-tubocurarine
and 5mM strychnine on the contractile protozoan Spirostomum and
found that both agents depress sensitivity to mechanical stimuli.
The effect of d-tubocurarine is obviously in agreement with our
data on Stentor while the effect of strychnine appears to be
different. However, strychnine at the concentration used by
Applewhite produces a depigmentation reaction in Stentor (Tartar,
1961) and elevates electrical shock thresholds. Therefore, a lower
concentration was used to generate the data reported above. At a
concentration of 0.03mM, we find strychnine does depress response
probability to mechanical stimuli. Since the electrical threshold
for contraction is also raised, we attribute this depression to a
non-specific effect and not to an action on mechanoreception.

The consistent effects on mechanoreception produced by nicotinic
cholinergic blocking agents suggests the involvement of a nicotinic
cholinergic receptor in protozoan mechanical stimulus transduction.
In accord with this hypothesis it has been observed that d-tubocurarine depresses the amplitude of mechanoreceptor potentials. However it is surprising that acetylcholine, carbachol, eserine,
neostigmine and acetylcholine plus eserine do not potentiate
Stentor's response probability to mechanical stimuli. Also
acetylcholine (1mM) and carbachol (1mM) injected directly onto the
animals do not elicit contractions as one might expect if acetylcholine is a depolarizing agent released by mechanical stimulation.
These negative results can be reconciled with the cholinergic receptor hypothesis in a variety of ways: 1) these agents may not

be able to permeate the animal's pellicle or membrane to reach the
receptor sites, 2) release of acetylcholine itself may not be
important in mechano-receptor potential production while the
cholinergic receptor is or 3) free acetylcholine may be normally
present in excess so that additional acetylcholine cannot potentiate the mechanoreception. A further description of the mechanisms
involved in mechanoreceptor potential production in protozoa cannot
be made at present. More data is needed to test some of the
hypotheses outlined above. However the possible involvement of a
cholinergic receptor mechanism suggests that analysis of this system
will continue to be both interesting and potentially relevant to
metazoan stimulus transduction.

In regard to the potential relevance of the analysis of protozoan transduction mechanisms, it is interesting that many vertebrate mechanoreceptors can be depolarized by acetylcholine even
though the receptor surface occurs on the afferent neuron and no
synaptic membrane is present (Brown and Gray, 1948; Douglas and
Gray, 1953; Jarrett, 1956). In these cases d-tubocurarine blocks
the response of these neurons to acetylcholine but does not affect
their sensitivity to mechanical stimuli. Therefore, it has been
concluded that cholinergic receptors are not involved in the transduction mechanisms of vertebrate mechanoreceptors (Gray, 1959).
However, Fitzgerald and Cooper (1971) have recently reported that
tactile sensitivity in the rabbit cornea is dependent on endogenous
acetylcholine concentration and have revived the suggestion that
acetylcholine may be important in sensory transduction. Thus the
elucidation of protozoan mechanoreceptor mechanisms may have
relevance to vertebrate mechanoreception, though the differing
effect of d-tubocurarine on these two preparations cautions against
any direct analogy.

TRANSDUCTION OF PHOTIC STIMULI
BY <u>STENTOR</u>

Phototaxis, the locomotor response of animals to photic stimuli,
has been studied extensively using blue-green algae, e.g., <u>Chlamydo-
monas</u>, and flagellate protozoa, e.g., <u>Euglena</u>, as subjects. This
literature also has been thoroughly reviewed (Halldal, 1964; Feinleib
and Curry, 1971). In all the cited studies the behavioral responses
of intact organisms have served as the dependent variable and
electrophysiological measurements have not been made. While considerable effort has been directed toward elucidating the chemical
nature of the photosensitive pigment in these organisms, action
spectra similar to the absorption spectra of known compounds have
been observed in only a few cases. For instance, the phototactic
action spectra of certain marine algae, <u>Platymonas subcordiformis</u>
in particular, is in agreement with the action spectrum of carotenoids

but to date it has proven impossible to extract the relevant pigment
from these organisms (Halldal, 1961). Similarly a flavin has been
suggested as the absorbent molecule producing phototaxis in _Euglena_
(Diehn and Kint, 1970), but its extraction has not been accomplished.
In most other cases the action spectra obtained from unicellular
organisms bears no clear relation to the absorption spectrum of
known pigments. This problem arises because of the complex structure
of the phototactic organisms. For instance _Euglena_ contains not
one but three pigment molecules: chlorophyll in chloroplasts,
carotenoids in the stigma, and the photosensitive pigment of the
small ($< 1\mu$) paraflagellar body. Phototaxis in these organisms is
thought to involve not only the absorption of light by one of these
pigments but also the screening of the photosensitive organ by the
other pigments. Therefore phototaxis in flagellates is a complex
behavioral event.

A photophobic response is exhibited by a few pigmented ciliates
of which _Stentor_ _coeruleus_ is both the largest and best known.
Sudden illumination of a swimming _Stentor_ results abruptly in
ciliary reversal and the other components of the "avoiding reaction"
(Mast, 1906). Illumination of a sessile animal whose posterior
is extended in the trumpet-shaped form elicits quite a different
response. In this case illumination produces an initial bending
and gradual shortening of the animal which may be followed after
10-40 seconds by a rapid contraction which appears similar to the
contractions produced by mechanical and electrical stimuli.

Electrophysiological recordings from intravacuolar electrodes
in _Stentor_ reveal that a graded receptor potential is produced by
sudden illumination of the animal (Wood, 1973). This receptor
potential can produce one of two subsequent electrophysiological
events depending on the configuration of the animal at the time of
stimulation. If the animal is extended then the receptor potential
may gradually increase in amplitude over the next 10-40 seconds and
an action potential may be elicited (Figure 3). These action
potentials are similar in form and amplitude to those produced by
mechanical stimuli and apparently are produced by the same mechanism
since they are correlated with the behavioral contraction as in the
case of action potentials elicited by mechanical stimuli (Wood, 1973).
However, if the _Stentor_ is in the "pear-shaped" form characteristic
of swimming animals, then illumination results in a receptor potential leading rapidly to a spike potential of variable amplitude which
is correlated with ciliary reversal, similar to the potentials seen
in _Paramecia_ (Naitoh and Eckert, 1969).

Illumination more reliably elicits ciliary reversal in swimming
Stentor than it does contractions in extended animals. We have
therefore utilized ciliary reversal as a dependent variable in determination of an action spectrum. Thresholds for elicitation of ciliary

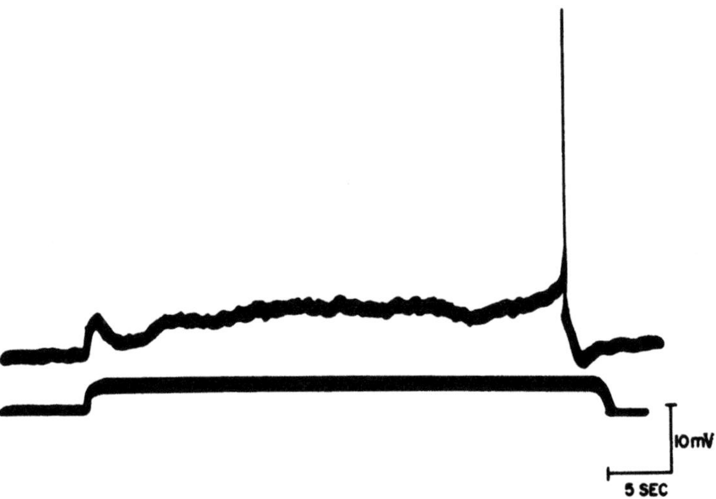

Figure 3. Upper trace shows a receptor potential which was produced by photic stimulation and recorded from a vacuole within the
animal. Thirty-seven seconds after initiation of the stimulation
an action potential and contraction was produced. The bottom trace
records the output of a monitoring photovoltaic cell.

reversal by photic stimuli of varying wavelength were determined
using the staircase technique (Cornsweet, 1962). The observed
action spectrum bears an unmistakeable resemblance to the _in situ_
absorption spectrum of the blue-green pigment of _Stentor_ as determined by microspectroscopy. Both spectra have a pronounced maximum
between 600 and 620 mµ and a depression between 520 and 540 mµ.
Thus, the blue-green pigment which has been called "stentorin"
(Lankester, 1873) and which gives _Stentor coeruleus_ its species
name appears to be the basis of its photosensitivity.

Nature has performed a genetic experiment which supports this
conclusion. Another species of _Stentor_, _Stentor niger_, is yellowish
brown in coloration rather than blue-green. It contains pigments
which absorb more blue light than red light (Barbier, Fauré-Fremiet
and Lederer, 1956). _Stentor niger_ is also more sensitive to blue
light than to red light (Tuffrau, 1957) which is the opposite
of _Stentor coeruleus_, which is more sensitive to red light. Thus the
change of pigments in _Stentor_ is associated with a corresponding
change in wavelength sensitivity.

Somewhat at variance with this conclusion is the observation
that _Stentor coeruleus_ depigmented by caffeine nevertheless produces
an "avoiding reaction" when illuminated (Tartar, 1972). However
our microspectroscopy measurements indicate that _Stentor_ which are

depigmented in this fashion and which appear white, nevertheless
exhibit reduced absorption peaks characteristic of the pigmented
animals. Thus, caffeine treatment may not produce complete depigmentation and hence only a difference in response threshold should
be expected between caffeine-treated and normal _Stentor_. We also
have observed such an elevation in response threshold in depigmented
animals.

Stentorin has been characterized as a meso-naphthodianthrone
compound on the basis of absorption data (Møller, 1960). It is
related to the photodynamic molecules hypericin and phagopyrin
found in plants and therefore is a unique photodynamic pigment
for an animal. The location of stentorin within the animal is
also rather unique, for it is located in 1μ pigment granules which
only have been found in _Stentor_ and in closely related protozoa.
While the function of these pigment granules is presently unknown,
they have been reported to contain cytochrome oxidase (Weisz, 1950)
and have been implicated in metabolic function (Tartar, 1961).

Absorption of light by stentorin appears to be the initial step
in production of the photic receptor potential and ultimately of the
observed behavioral responses. Although the photic receptor potential has the same polarity as the mechanoreceptor potential it is
produced by a different mechanism. This independence of mechanical
and photic receptor mechanisms is demonstrated by the failure of
d-tubocurarine and other nicotinic blocking agents to depress
sensitivity to photic stimuli while markedly depressing sensitivity
to mechanical stimuli. Nor does depression of mechanical receptor
potentials by repeated mechanical stimulation (Wood, 1971) effect
a depression in the animal's sensitivity to photic stimuli (Wood,
1973).

A clue to the mechanism generating photic receptor potentials
can be found in their time course. The onset of these potentials
occurs with a latency of 0.1-0.3 seconds after the initiation of
illumination; therefore, they do not have the rapid onset which
characterizes the conductance change mechanism producing mechanical
receptor potentials. Once initiated these potentials often show
5-15 mv. waves during the course of illumination. This result
suggests that some form of oscillatory process is activated by
illumination which is probably related to a metabolic function.

We have found that membrane conductance as measured with rectangular current pulses does not change during the photic receptor
potential. However, the intravacuolar recording site was such that
small changes in external membrane conductance might not have been
measureable. Also shifts in intracellular potential produced by
injected direct current did not change the polarity or amplitude
of the photic receptor potential but did reverse the polarity of

the mechanical receptor potential (Wood, 1973). Thus I conclude
that the photic receptor potential is not produced by changes in
membrane conductance but rather is produced by stimulation of an
electrogenic pump. This conclusion is strengthened by the presence
of the absorbing pigment in granules which contain at least one
enzyme involved in the electron transport chain of oxidative phosphorylation.

RECONSIDERATION ON THE USE OF PROTOZOA
AS MODELS OF STIMULUS TRANSDUCTION

At the outset of this paper I noted that certain neural and
receptor mechanisms appear in organisms positioned along many
different branches of the phylogenetic tree. It was then assumed
that these organisms exhibited similar physiological and biochemical characteristics because they had certain genetic similarities which they derived from a common primordial ancestor. By
extrapolation it was argued that protozoa may share some of this
gene pool and hence possess some interesting "neural" mechanisms.
In reviewing protozoan stimulus transduction mechanisms we find
exciting parallels between metazoan and protozoan mechanoreceptor
mechanisms. Both classes of organism appear to employ membrane
conductance changes to alter ionic currents through the membrane
and thus generate receptor potentials. In both cases these receptor
potentials may serve to trigger transient potentials which mediate
the eventual behavioral response. The sensitivity of protozoan
mechanoreceptors to cholinergic blocking agents suggests a possible
cholinergic system in protozoan mechanoreception. The involvement
of a cholinergic system in metazoan mechanoreception also has been
suggested. Thus the original hypothesis that protozoa may serve
as appropriate models for the study of sensory transduction mechanisms still appears valid with regard to mechanoreception.

However photoreception in _Stentor_ is unusual among animals
since it involves a class of pigments otherwise observed only in
plants, and has a receptor potential mechanism which appears independent of membrane conductance changes. Photoreception in _Stentor_
appears to have followed a different evolutionary line from that
taken by metazoan photoreceptors.

Where then does this leave the original hypothesis of similarity
between protozoan and metazoan transduction mechanisms? Both the
membrane conductance changes observed in protozoan mechanoreception
and the carotenoid pigments absent from _Stentor_ photoreception are
found throughout the phylogenetic scale. Therefore the argument
that physiological and biochemical mechanisms found in animals from
many branches of the phylogenetic tree also will be found in protozoa
cannot be universally accepted.

The most obvious difference between mechanoreception and photoreception in protozoa is in the ubiquity of these processes among protozoa themselves. Sensitivity to mechanical stimuli can be observed among ciliates, flagellates and rhizopods and thus is exhibited throughout the protozoan phylum. On the other hand photosensitivity occurs only in certain classes of protozoa, and among ciliates only in _Stentor_ and in a few closely related species. Therefore, on the basis of these examples it seems appropriate to use protozoa as model systems if the mechanisms under study are not only widely distributed throughout the phylogenetic scale but also widely distributed throughout the protozoa.

REFERENCES

Applewhite, P.B. 1972. Drugs affecting sensitivity to stimuli in the plant mimosa and the protozoan _Spirostomum_. _Physiol. Behav._, _9_, 869-871.

Barbier, M., Fauré-Fremiet, E. and Lederer, E. 1956. Sur les pigments du Cilie' _Stentor niger_. _C.R. Acad. Sci._, _Paris_, _242_, 2182-2184.

Brown, G.L. and Gray, J.A.B. 1948. Some effects of nicotine-like substances and their relation to sensory nerve endings. _J. Physiol._, _107_, 306-317.

Cornsweet, T. 1962. The staircase-method in psychophysics. _Amer. J. Psychol._, _75_, 485-491.

Diehn, B. and Kint, B. 1970. The flavin nature of the photoreceptor molecule for phototaxis in _Euglena_. _Physiol. Chem. Phys._, _2_, 483-488.

Douglas, W.W. and Gray, J.A.B. 1953. The excitant action of acetylcholine and other substances on cutaneous sensory pathways and its prevention by hexamethonium and d-tubocurarine. _J. Physiol._, _119_, 118-128.

Eckert, R. 1965a. Bioelectric control of bioluminescence in the dinoflagellate _Noctiluca_. _Science_, _147_, 1140-1142.

Eckert, R. 1965b. Asynchronous flash initiation by a propagated triggering potential. _Science_, _147_, 1142-1145.

Eckert, R. 1972. Bioelectric control of ciliary activity. _Science_, _176_, 473-481.

Eckert, R. and Naitoh, Y. 1972. Bioelectric control of locomotion in the ciliates. _J. Protozool._, _19_, 237-243.

Eckert, R., Naitoh, Y. and Friedman, K. 1972. Sensory mechanisms in _Paramecium_. I. Two components of the electric response to mechanical stimulation of the anterior surface. _J. Exp. Biol._, _56_, 683-694.

Feinleib, M.E. and Curry, G.M. 1971. The nature of the photoreceptor in phototaxis. In _Handbook of Sensory Physiology_. I., W.R. Lowenstein (Ed.), New York: Springer-Verlag. pp. 366-395.

Fitzgerald, G. and Cooper, J. 1971. Acetylcholine as a possible
    sensory mediator in rabbit corneal epithelium. <u>Biochem</u>. <u>Pharm</u>.,
    <u>20</u>, 2006–2007.
Frankenhaeuser, B. 1960. Quantitative description of sodium
    currents in myelinated nerve fibers of <u>Xenopeis laevis</u>. <u>J</u>.
    <u>Physiol</u>., <u>131</u>, 341–376.
Ginsborg, B.L. 1967. Ion movements in junctional transmission.
    <u>Pharmac</u>. <u>Rev</u>., <u>19</u>, 289–316.
Gray, J.A.B. 1959. Initiation of impulses at receptors. In
    <u>Handbook of Physiology</u>, Part I, I., Neurophysiology, J. Field
    (Ed.), Washington: American Physiological Society. pp. 123–145.
Halldal, P. 1961. Ultraviolet action spectrum of positive and
    negative phototaxis in <u>Platymonas subcordiformis</u>. <u>Physiol</u>.
    <u>Plantarum</u>, <u>14</u>, 133–139.
Halldal, P. 1964. Phototaxis in protozoa. In <u>Biochemistry and
    Physiology of Protozoa</u>, Vol. 3, S.H. Hatner (Ed.), New York:
    Academic Press. pp. 277–296.
Hodgkin, A.L. and Huxley, A.F. 1952. A quantitative description
    of membrane current and its application to conduction and excita-
    tion in nerve. <u>J</u>. <u>Physiol</u>., <u>117</u>, 500–544.
Jarrett, A.S. 1956. The effect of acetylcholine on touch receptors
    in frog's skin. <u>J</u>. <u>Physiol</u>., <u>133</u>, 243–254.
Jennings, H.S. 1906. <u>Behavior of the Lower Organisms</u>. New York:
    Columbia University Press.
Lankester, E.R. 1873. Blue stentorin, the coloring matter of
    <u>Stentor coeruleus</u>.. <u>Quart</u>. <u>J</u>. <u>Micros</u>. <u>Sci</u>., <u>13</u>, 139–142.
Mast, S.O. 1906. Light reactions in lower organisms. I. <u>Stentor
    coeruleus</u>. <u>J</u>. <u>Exp</u>. <u>Zool</u>., <u>3</u>, 359–399.
Møller, K.M. 1960. On the nature of Stentorin. <u>Compt</u>. <u>rend</u>.
    <u>trav</u>. <u>Lab</u>. <u>Carlsberg</u>., <u>32</u>, 471–497.
Naitoh, Y. and Eckert, R. 1969. Ionic mechanisms controlling be-
    havioral responses of <u>Paramecium</u> to mechanical stimulation.
    <u>Science</u>, <u>164</u>, 963–965.
Naitoh, Y. and Eckert, R. 1973. Sensory mechanisms in <u>Paramecium</u>.
    II. Ionic basis of the hyperpolarizing receptor potential.
    <u>J</u>. <u>Exp</u>. <u>Biol</u>., <u>59</u>, 53–65.
Tartar, V. 1961. <u>The Biology of Stentor</u>. New York: Pergamon Press.
Tartar, V. 1972. Caffeine bleaching of <u>Stentor coeruleus</u>. <u>J</u>. <u>Exp</u>.
    <u>Zool</u>., <u>181</u>, 245–252.
Tuffrau, M. 1957. Les Facteurs essentiels du phototropisme chez
    le Cilié heterotriche <u>Stentor niger</u>. <u>Bull</u>. <u>Soc</u>. <u>Zool</u>., <u>France</u>,
    <u>82</u>, 354–356.
Weisz, P.B. 1950. On the mitochondrial nature of the pigmented
    granules in <u>Stentor</u> and <u>Blepharisma</u>. <u>J</u>. <u>Morph</u>., <u>86</u>, 177–184.
Wood, D.C. 1970a. Parametric studies of the response decrement
    produced by mechanical stimuli in the protozoan, <u>Stentor coeruleus</u>.
    <u>J</u>. <u>Neurobiol</u>., <u>1</u>, 345–360.

Wood, D.C. 1970b. Electrophysiological studies of the protozoan,
  Stentor coeruleus. J. Neurobiol., 1, 363-377.
Wood, D.C. 1971. Electrophysiological correlates of the response
  decrement produced by mechanical stimuli in the protozoan, Stentor
  coeruleus. J. Neurobiol., 2, 1-11.
Wood, D.C. 1972. Generalization of habituation between different
  receptor surfaces of Stentor. Physiol. Behav., 9, 161-165.
Wood, D.C. 1973. Stimulus-specific habituation in a protozoan.
  Physiol. Behav., 11, 349-354.
Wood, D.C. 1974. Ionic mechanisms in the generation of action and
  receptor potentials in Stentor. (In preparation).

GENETIC DISSECTION - AN APPROACH TO NEUROBIOLOGY

Ching Kung
Department of Biological Sciences
University of California
Santa Barbara, California 93106

This chapter serves two purposes: First, to show how genetics can be used as a tool in neurobiological research and, secondly, to argue that bacteria and protozoa can be used as model systems for such an interdisciplinary approach.

A GENETIC APPROACH TO NEUROBIOLOGY

Suppose that we are given a TV set in perfect condition but without an instruction manual or circuit diagrams, and we are asked to figure out how it works. Depending on one's temperament and training, many approaches may be taken. Some people will simply turn the set on, shelter it from any disturbance and carefully observe what comes out of it, from Sesame Street to Johnny Carson. Some will twist all the knobs in front and put in cables from the back and see what such manipulations do to the picture. Some will proceed to open the chassis and examine the interior, noting that x number of parts are connected by y number of wires with a certain configuration. Others may feel like measuring voltage and impedence across various elements. Then, there are those who will smash the box, extracting various elements and finding that a TV set has 0.003% of tungsten or that set A has significantly more solder than set B. If the TV set is the CNS and the picture, behavior, I have just offended the ethologists, the psychologists, the neuroanatomists, the neurophysiologists and the neurochemists by grossly distorting what they are trying to do.

The fact that all these experts from diverse disciplines concentrate their talents here is a reflection of the complex structures and the sophisticated functions of the CNS. The practical importance

and the intellectual appeal of explaining the physical basis of the
mind attract scholars from even wider realms. The current exodus
from classical molecular biology to neurobiology attests to this.
As a geneticist affected some years ago by the magnet of neurobiology, I wondered if my own discipline could contribute to the
search for the molecular mechanisms of behavior. There was a
behavioral genetics field, but it was mainly a branch of genetics
concerning itself with the patterns of inheritance of behavioral
phenotypes. That genetics could indeed be used as an experimental
tool in the study of behavior was first pointed out by Professor
Seymour Benzer (1967):

> Complex as it is, much of the vast network of cellular
> functions has been successfully dissected, on a micro-
> scopic scale, by the use of mutants in which one element
> is altered at a time. A similar approach may be fruitful
> in tackling the complex structures and events underlying
> behavior, using behavioral mutations to indicate modifica-
> tions of the nervous system.

Returning to our TV analogy, what Benzer has suggested and what
we generally called "genetic dissection" is a procedure much like
a kind of "tube testing". That is to say, one way of trying to find
out how the TV works is to take out one tube, one transistor, one
wire, etc., at a time and see what effect is produced on the picture.
The defects induced often enable one to obtain information on the
functions of each of the elements.

Bungling up a biological system by deleting the functioning
elements one at a time should not be an alien experimental procedure
to the psychologists or the physiologists. The familiar use of
pharmacological agents in the investigation of the nervous system
does just that. Drugs are applied to block a specific channel, to
bind to a specific receptor site or to inactivate a specific pump.
The information we have about the density of Na-channels, the
function of Na-pumps and the isolation of ACh-receptors was gained
through the use of tetrodotoxin, ouabain and α-bungarotoxin
respectively. The genetic approach does not differ from the pharmacological approach in principle. All proteins (and thus, all the
enzyme catalyzed reactions) are under genic control, and therefore
open to mutational attack. The genetic approach can thus be taken
systematically, at least in theory, and does not have the drawbacks
of the pharmacological approach. Drugs and toxins are discovered
haphazardly and limited in number.

A general program for genetic dissection of the mechanisms of
behavior may be described as follows:

(1) <u>Induction of Mutations</u>. Mutations are alterations of the
genetic information (primary structure of DNA) that lead to

corresponding changes in the gene products (RNA or proteins).
Mutations that occur spontaneously can be encouraged by treatment
with various mutagens whose chemical actions are often known.
Although the chemistry prescribes that the mutagen dosage be pro-
portionate to the number of mutational hits, in practice, over-
dosing a population only causes death or genetic death (lethal
mutations) and complicates the system with multiple mutants. Thus,
for each gene even the induced mutation rate remains low.

(2) _Mutant Selection_. Mutagenesis is not a process that can
be directed to hit specific genes, even if one knows where they
are located on the genetic map. All genes and therefore all gene
products have a certain probability of being mutated after mutagen
treatment, although the probabilities of different genes may be
different. In practice, mutants are not recognized by the
difference in their DNA but by their phenotypic expressions which
are the results of the non-functioning or malfunctioning of the
altered RNAs or proteins. After successful mutagenesis, one can
scan the mutagenized populations and find a large variety of
mutants. Their mutations include various enzyme defects, morphological alterations, developmental aberrations, and behavioral
peculiarities. Since the mutation rate of any specific gene remains
low even after optimal mutagen treatment, one needs to devise certain
screening procedures to select for the mutants. The methods of
selection vary with the phenotypes of interest. In our case,
"misbehavior" would furnish the key for picking out mutants with
defects in the generation of that specific behavior.

(3) _Analyses of phenotypes_. The manifestation of the genetic
defects, although ultimately behavioral, must nevertheless relate
to structural differences in the molecules and all the biochemical
reactions and physiological processes involved. Below the level of
behavior such defects may surface as gross or fine structural alterations of specific sites, as developmental difficulties at
specific times of the life cycle, or as various physiological ab-
normalities manifested in various tissues, cells or organelles.
The study of the phenotypes will thus go beyond formal genetics.
By collating the phenotypic expressions detected by different
structural and functional analyses, we may gain insight into the
basis of the behavioral defects. Such knowledge often permits one
to make inferences about the normal mechanisms of behavior in the
wild type.

(4) _Identification of the gene products_. In molecular terms,
the difference between the wild type and a mutant is often minimal,
i.e., only one gene product difference, although the gene product
may have multiple functional effects and expressions (pleiotropism).
Such a difference provides a clue which can be used to track down
the molecular species of special interest. Unless a molecular

species has a certain enzyme activity or has unique binding properties it cannot be tagged and recognized. Molecules important in
neurobiology, such as various membrane proteins, may not be identifiable this way. Yet, relevant mutations can alter or delete these
proteins, so that by comparing the molecular profile of the wild
type and the mutants we can identify them. Once isolated and
identified, the composition, structure and activity of these
molecules can be studied by existing biophysical and biochemical
techniques.

UNICELLS AS MODEL SYSTEMS

    Biologists are accustomed to the use of model systems which
offer experimental advantages. Thus, the embryologists have their
frogs and sea urchins, the geneticists have their _Drosophila_ and
Escherichia coli. Perhaps we should be reminded that, after all,
the giant squid axon and the frog sartorius are also model systems
for excitable tissue in general. It is therefore not very far-
fetched to choose an insect (Benzer, 1967) or a nematode (Brenner,
1973) as a model system for the genetic dissection of the nervous
system, because of the ease with which these animals can be manipulated genetically. I would argue further that even bacteria
and protozoa can be regarded as model systems for certain neurophysiological investigations.

    Ciliated protozoa were once thought to have a prototype of a
nervous system. This was mainly due to the widely quoted experiments by Taylor (1920) in which he found that microsurgical severing
of the subcellular fibers in _Euplotes_ interrupted the coordination
in the movement of the motile organelles (the membranelles and the
cirri, both compound ciliary structures). Some of Taylor's observations were not repeated in cells of better post-surgical recovery
and most of his interpretations have now been questioned by recent
investigations (Okajima and Kinosita, 1966; Naitoh and Eckert, 1969).
However, the notion that a primitive nervous system (the neuromotor
fibers and the "motorium") presides in a "primitive" organism
(_proto_-zoan) is so appealing, as is evident in the secondary accounts
of Taylor's experiments, that it will take time to correct this
impression. It all amounts to what I call a "metazoopomorphic view"
of protozoa much like the anthropomorphic view of animals, a mistake
that is almost unavoidable given the status of _proto_-zoology.

    Since I have segregated nerveless protozoa from metazoa, or
more generally, the unicellular from the multicellular organisms,
the question of why protozoa and bacteria in neurobiology is even
more pointed. These organisms are often referred to as aneural.
To my mind, a protozoan (or maybe even a bacterium) is about as
aneural as it is acellular. One adopts a unicellular or an acellular

view of protozoa depending on one s definition of a cell. Protozoa
are aneural only in the sense that they do not have a nervous
system. They nevertheless are much like neurons or sense cells.

That the basic properties of excitable cells can be found in
some unicellular organisms is best illustrated in the ciliated
protozoan Paramecium.

Jennings (1906) used Paramecium as a model in developing his
concept of an "action system" in the behavior of lower organisms.
This system in Paramecium consists mainly of the avoiding reaction.
This reaction interrupts the usual forward helical swimming with
backward movement, which is accomplished by reversal of the ciliary
beat. Taxes or kineses induced by chemicals, heat or mechanical
disturbances are accomplished by this relatively stereotyped reaction,
often repeatedly. The electrophysiological correlates of the behavior are now fairly well understood. Paramecium was among the
first cells to be studied with intracellular microelectrodes
(Kamada, 1934). The membrane is inside negative at rest, as are
most other cells. The resting potential level depends on the concentration of many cations in the bath as predicted for a poorly
selective membrane by the Goldman relation (Naitoh and Eckert,
1968a). Membrane bound Ca^{++} is apparently directly related to
membrane resistance. Kinosita et al. (1964a,b) have shown that
ciliary reversal, hence backing during avoiding reactions, is
correlated with membrane depolarization. The reversal of ciliary
beat upon depolarization is now thought to be the result of influx
of Ca^{++} due to Ca^{++} activation (Eckert, 1972). The voltage sensitive change in Ca^{++} conductance can become regenerative in generating action potentials upon proper electrical, mechanical or ionic
stimulation. The spikes are usually graded to the strength of the
stimuli. In Ba-Ca solutions, however, all-or-none action potentials
appear spontaneously and can also be evoked by outward current
(Kinosita and Murakami, 1965; Naitoh and Eckert, 1968b). In reaction
to mechanical stimuli at the anterior end, receptor potentials can
be generated which are readily discernable from the active electrogenesis they trigger (Naitoh and Eckert, 1969; Eckert, Naitoh and
Friedman, 1972). It is thus evident that the paramecium membrane
has some of the basic properties of excitable cells and sense cells.

If we accept the idea that some protozoa or bacteria may be
comparable to the excitable cells in higher forms in terms of basic
molecular mechanisms, we can argue further that they are not just
marginally acceptable model systems but, in fact, desirable systems
for the purpose of genetic dissection of molecular events important
in neurobiology. There are great advantages in using these lower
forms as experimental material. I will attempt to list the major
ones:

(1) These organisms can be grown rapidly into huge populations
without much effort or expense. This fact facilitates both genetic
and biochemical analyses, in which the size of the population or
the rapidly available biomass are often the limiting factors. We
save time, space and energy working with bacteria instead of
elephants, if bacteria can serve the purpose.

(2) Chemical analysis of the nervous system is greatly
hampered by the presence of many cell types lying side by side in
the same piece of tissue. It is often difficult to ascertain
whether certain molecular species are present or are present in
abundance in the neurons of interest or in the other neurons, glial
cells or the connective tissue. All uni-cells have asexual repro-
ductive cycles which produce clones of cells of the same genotype
and same phenotype. Interpretation of the biochemical findings of
such a uniform population of cells is much easier and more meaningful.

(3) There is a spectrum of molecular genetic techniques and
biochemical methods available in the study of lower forms, especially
bacteria, as shown by the current literature. Although in principle
such methods could be used in higher forms, in practice they have
not been adopted successfully or effectively, partly due to the
complexity of the organisms and partly due to the lack of proper
tools such as transducing phages and suppressor mutant strains.

(4) If one induces mutations in a multicellular organism, it
is often difficult to ascertain that certain cell groups are the
primary targets of mutations although the final expressions are
behavioral. In unicellular research, when one deals with only one
kind of cells, such problems simply do not arise.

(5) Cultured cells from nervous systems of higher forms, such
as neuroblastoma, have some of the advantages mentioned above.
However, there is no nuclear reorganization by sexual reproduction
in cultured somatic cells. Formal genetic analyses cannot be performed in such cells. Free-living bacteria and protozoa are
capable of conjugations and transfer of genetic material from one
cell to another. Microbial genetics is certainly well advanced.

However, the choice of using bacteria or protozoa as model
systems in neurobiological research does, of course, have a few
drawbacks. These cells are, at best, analogs of excitable cells and
not models of the nervous system. Some of the interesting problems
in neurobiology are not at the subcellular or molecular levels.
The problem of cell to cell communication through synapses or the
problem of precise and predestined wiring patterns of the CNS
cannot be studied in bacteria or protozoa. Even if one confines
himself to subcellular and molecular problems, there is still one
major reservation that should be kept in mind. These entities are

not just cells, they are organisms at the same time. Unlike the
neurons which have evolved to become specialized for a unique
function, bacteria or protozoa cannot afford to do just that.
Because of the simultaneous presence of many biological processes,
interpretations of observations of any specific function should be
especially cautious.

GENETIC ANALYSES OF CHEMOTAXIS IN ESCHERICHIA COLI

Although the electrophysiological study of procaryotes has
yet to be developed due to their small size, behavior of motile
bacteria has been carefully observed. That bacteria can be
attracted or repelled by various chemicals is an old observation
(see Jennings, 1906). A motile bacterium tumbles periodically,
resulting in changes in its course of movement. With increasing
concentrations of an attractant, the frequency of tumbles of the
bacterium decreases, resulting in a higher probability of its
staying close to the attractant. A higher frequency of tumbles,
and, therefore, a higher rate of turns, occur when the cell is
moving toward a lower concentration of the attractant or a higher
concentration of a repellant. A bacterium apparently detects a
chemical gradient not by differentiating the gradient at various
points of its body but by distinguishing the chemical concentrations monitored at different times during the course of its movement
(MacNab and Koshland, 1972).

The modern attack on the problem of chemotaxis in bacteria
has been led by Adler (1969) in an effort to understand the molecular mechanisms of this piece of behavior. A priori, there could
be three components necessary for chemotaxis. They are the
chemical detection component, the locomotor apparatus and the
connecting element between the two. The conceptually distinguish-
able components may not necessarily have their molecular counter-
parts, e.g., one can imagine that the motile apparatuses may have
a sensory function as well or that the two are directly coupled
with no intermediates. Adler and his colleagues, however, have
provided ample evidence that there are indeed three essential
components, each of different molecular species, for chemoreception,
motility and the coupling of the two. Although ultrastructural,
physiological and biochemical analyses were important, much of their
work has been accomplished through the use of genetic mutants. In
line with the idea of genetic dissection discussed above, the presence and the function of a molecular component important in chemotaxis was demonstrated by obtaining a class of mutants that failed
to perform chemotaxis due to the loss of that molecular species.
I shall attempt to sketch their work on the three components of
chemotaxis.

(1) <u>Motility</u>. Destruction of the mechanisms associated with
motility by mutations leads to nonmotile mutants. Two groups of
such mutants have been identified. First, there are mutants with
no flagella or those with abnormal flagellar morphology. In these
mutants the genes specifying the flagellar protein (flagellin) or
the genes governing its synthesis are mutated. In addition, there
are mutations that affect the function but not the morphology of
the flagella. The phenotypic expression is paralysis of the
bacteria with apparently normal flagella (Armstrong and Adler,
1967). The relevant gene products may be those needed to channel
the energy to the flagella and allow it to beat, although how the
beat is energized is still not clear.

(2) <u>Chemoreception</u>. The presence of specific mechanisms for
detecting chemicals is demonstrated by obtaining mutants with
specific chemotactic failures (Adler, 1969; Adler, Hazelbauer and
Dahl, 1973; Mesibov and Adler, 1972). Mutants have been found
that are no longer attracted by a specific group of chemicals
(e.g., aspartate and glutamate), but have normal motility and,
more important, can still be attracted by other chemicals. Mutants
with specific loss of chemoreception of repellants also have been
found (Tso and Adler, 1974). One can ascertain the number of
different kinds of chemoreceptors by finding mutants that fail
specific chemotactic tests one at a time. In <u>E</u>. <u>coli</u>, for instance,
different receptors for galactose, glucose, ribose, aspartate,
serine, etc., can be delineated by mutations. The mutant approach
also has been used to separate the events of detection of a chemical
from the events of its permeation and metabolism. Mutants defective
in galactose transport, as well as constructed triple mutants that
no longer metabolize galactose, can nevertheless be attracted by
galactose through chemotaxis (Adler, 1969).

(3) <u>The coupling elements</u>. The elements that channel the
information from the chemoreceptors to the flagella can also be
subjected to the scalpel of genetics. Armstrong, Adler and Dahl
(1967) searched for such mutants and have isolated 40 of them.
These mutants are generally nonchemotactic, unable to be attracted
by all the attractants, as one would expect. Yet they are not
nonflagellated or paralyzed. Complementation tests by abortive
transduction with bacteriophage P1 showed that 38 of the nonchemotactic mutations tested belonged to three complementation
groups (Armstrong and Adler, 1969). With a few exceptions, functional complementation of two mutations implies that the genes
involved direct the synthesis of two discretely separated proteins.
The finding of three genes (cistrons) all leading to general non-
chemotaxis indicates that at least three structural components are
present on the common pathway between various chemoreceptor molecules and the flagella.

In <u>E</u>. <u>coli</u> one can, in fact, isolate the relevant gene products
and study them <u>in</u> <u>vitro</u>. A galactose binding protein, for example,
was isolated which is apparently a receptor for galactose taxis
(Hazelbauer and Adler, 1971).

GENETIC DISSECTION OF THE MEMBRANE FUNCTIONS IN <u>PARAMECIUM</u> <u>AURELIA</u>

<u>Paramecium</u> <u>aurelia</u>, has two distinct advantages as experimen-
tal material for genetics and neurophysiology. Unlike bacteria,
it is a big cell, measuring on the average 150 μ in length. This
allows intracellular recording, which is important in understanding
membrane biophysics in orthodox electrophysiological terms. Unlike
other diploid organisms, <u>P</u>. <u>aurelia</u> is capable of extreme "inbreeding"
through the process known as autogamy (Sonneborn, 1970). During
autogamy, two identical gametic nuclei (resulting from a mitosis
of one meiotic product) fuse to restore the diploid state. Since
the two haploid gametic nuclei are identical, the diploid nucleus
formed will be completely homozygous at all genic loci. Such a
complete homozygote is like a haploid organism because all genes,
regardless of whether they are recessive or dominant, will be
expressed. Mutations, most of them recessive, are not easily
detected in most diploid organisms. In <u>Paramecium</u>, the easily
controllable process of autogamy circumvents this problem and makes
it rather simple to identify and select for various mutations.

Taking advantage of the ease of genetic manipulations and the
practicability of intracellular recording, genetic dissection of
the excitable membrane of <u>P</u>. <u>aurelia</u> was carried out following the
procedure outlined in the first section of this chapter.

We first mutagenized populations of <u>Paramecia</u> with N-methyl-N'-
nitro-N-nitrosoguanidine and then induced autogamy in them by
moderate starvation. After a few exautogamous fissions (a period
known as the phenomic lag), the expanded populations were screened
for behavioral mutants. In some cases, a selection procedure was
devised to obtain fractions greatly enriched with the mutants
desired. The mutants selected were then studied behaviorally
(Kung, 1971a; Chang and Kung, 1973a), genetically (Kung, 1971b;
Chang and Kung, 1973b) and electrophysiologically (Kung and Eckert,
1972; Satow and Kung, 1974; Satow, Chang and Kung, 1974). Behavioral
mutants with membrane defects obtained to date are of the following
types:

(1) <u>Pawns</u>. The outstanding characteristic of these mutants
is the lack of ability to reverse their ciliary beats, even upon
strong chemical and mechanical stimulation. The mutants always
swim forward, hence, the name (Figure 1). One hundred and fifty
lines of Pawns from several mutagenesis experiments are now
available. Those that have been analyzed genetically show that the

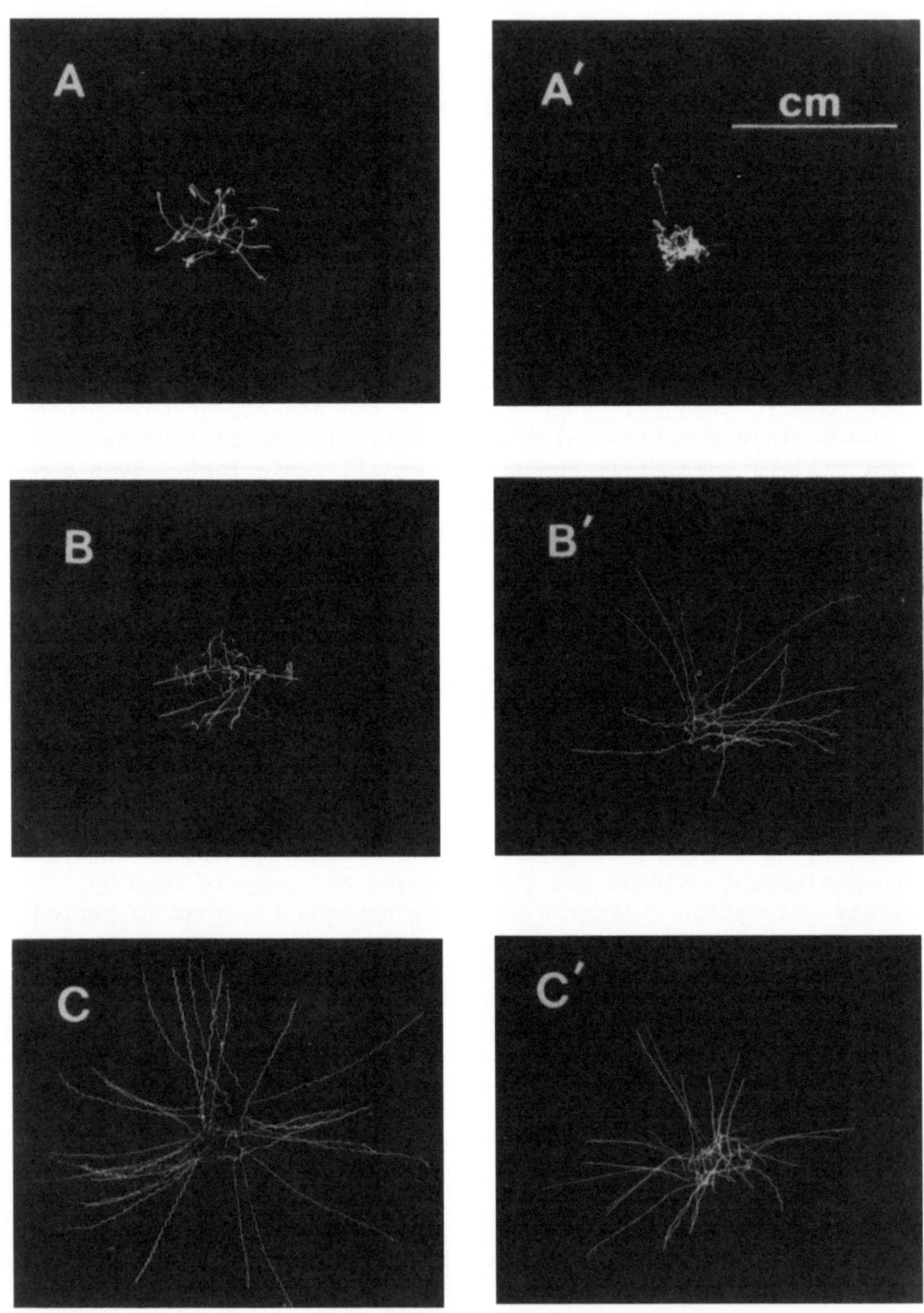

Figure 1. Behavioral responses to Ba^{++} solution of three strains of P. aurelia. These are dark field photomacrographs taken at $23 \pm 1°C$ in which the continuous lines record the movement of cells during the 13.3 ± 0.1 seconds immediately after they were put into the Ba^{++} solution. A,A'; wild type (strain 51s). B,B'; a ts Pawn mutant (strain d4-133). C,C'; a temperature independent Pawn mutant (strain d4-95). A,B,C; cells grown at 23°C. A',B',C'; cells were grown at 35°C and were photographed within 30 minutes after they were taken from the 35° C incubation. A and A' show wild type repeatedly avoiding the Ba^{++} solution. This reaction results in confining the cells to the vicinity of the area in which they were put. C and C' show the unconditional Pawn mutants swimming into the surrounding Ba^{++} solution along their usual helical courses with no avoiding reaction. This results in the sun-ray patterns. B and B' show the ts Pawn mutants behaving like the wild type when grown at 23° C and like Pawns when cultured at 35° C.

phenotype is controlled by recessive alleles on three unlinked
loci (Kung, 1971a; Chang and Kung, 1973b; Chang et al., 1974).
We (Kung and Eckert, 1972) have analyzed the membrane of one of
the Pawns (Pawn d4-95) through intracellular recording. While
there is no observable change in the passive properties of the
membrane in this Pawn, there is a dramatic loss of action potentials.
Upon proper electric and/or ionic stimulation, the wild type P.
aurelia fires off as expected but the Pawn simply reacts electro-
tonically (Figure 2). The possibility of K^+ leakage created by
the mutation is ruled out, since the input resistance of the wild
type and the mutant appears to be the same. A likely interpretation is that the voltage sensitive Ca^{++} gate, an important link in
the positive feedback system leading to the upstroke of the action
potentials, is wrecked by the mutation. The view that the mutational lesion is confined to the membrane was reinforced by Kung
and Naitoh (1973) with triton X-100 extracted models. The cilia
of these tritonated models of Pawns can be reactivated to beat
by Mg^{++} and ATP and to reverse by Ca^{++} and ATP as the cilia of the
wild type models. This indicates that the apparatuses needed to
beat and to reverse the beat are intact in this mutant. We now
have evidence that other Pawns are defective in their membranes
in a similar manner (Satow and Kung, 1974).

(2) Temperature sensitive Pawns. Five lines bearing genes on
two loci and exhibiting a temperature dependent Pawn character
have recently been selected and identified (Chang and Kung, 1973a,b).
These cells are normal when grown at 23°C and are Pawns at 35°C
(Figure 1). The membranes of these mutants are excitable at the
permissive temperatures but not at the restrictive ones (Satow,
Chang and Kung, 1974). The dependence of the kinetics of phenotypic alteration upon temperature step jumps indicates that the
gene products are probably proteins involved in the synthesis or
organization of the membrane Ca^{++} gate. Heat sensitive phenotypes
are generally understood as the result of temperature dependent
macromolecular alterations, most likely heat dependent protein
conformational changes. Aside from their possible implications as
to the molecular mechanisms of membrane excitation, the temperature
sensitive Pawns are interesting material for the study of membrane
function since the normal process can now be disrupted and restored
in the system at will with the change of presumably only one molecular species.

(3) Fast-2. Behaviorally, these mutants are highly Na^+-
insensitive, i.e., they do not avoid high concentrations of Na^+,
as the wild type does by way of ciliary reversal. Electrophysiological analysis of this mutant showed a loss of depolarizing spikes
upon introduction of Na^+ into the bath (Satow and Kung, 1974).
The resting potential level of the Fast-2 membrane becomes more
negative when Na^+ appears in the bath. The abnormal relation of

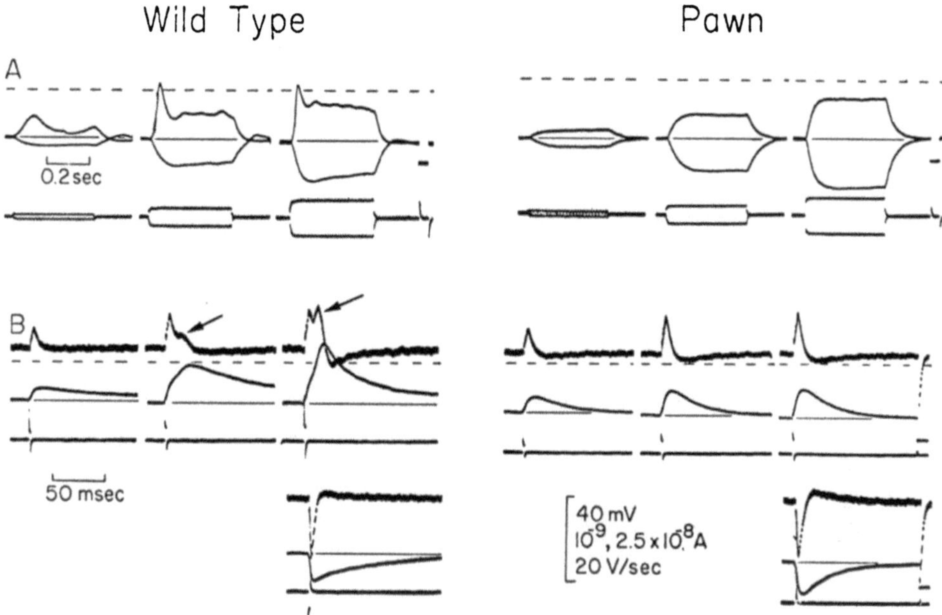

Figure 2.   Electrical excitability of the behavioral wild type
(d4-85).   P. aurelia and mutant (d4-95) tested in 1 mM $CaCl_2$, 4 mM
$BaCl_2$ and 1 mM Tris-HCl.   Wild type (left) shows characteristic
all-or-none barium action potentials in response to long- (A) and
short- (B) stimulus pulses, and as after-discharges and spontaneously
(C).   The critical threshold for the barium spike produced by wild
type is evident in series B, in which the same amount of current
(4 nA, 2 msec) produced no active responses in the 2nd frame
and a full action potential in the 3rd frame.   Pawn (right) shows
no signs of local active responses or action potentials.   Dashed
lines indicated the reference (Vm = 0) potential.   <u>Upper trace is
Vm and lower trace I, in A and C; upper trace is dVm/dt and lower
trace Vm in B.</u>

membrane potential and cationic concentrations leads us to conclude
that Fast-2 is a K^+-permeability mutant (Satow and Kung, in preparation).

     (4) <u>Paranoiacs</u>.   These mutants are characterized by sporadic
and apparently spontaneous violent avoiding reactions in regular
culture medium.   This characteristic appears to be Na^+ related
(Kung, 1970).   These mutants over-react when they encounter relatively high Na^+ in the medium.   Instead of a transient ciliary
reversal, prolonged backward swimming is observed.   Intracellular
recording showed that when Na^+ flows into the bath, action potentials with a greatly prolonged plateau phase occur instead of the

typical spikes of the wild type in an identical situation (Satow
and Kung, 1974). Five lines, at 4 unlinked loci, with this peculiar
phenotype have recently been discovered (Chang and Kung, unpublished).
Paranoiacs are the only behavioral mutants in which the mutant
alleles are either co-dominant or dominant to the wild type allele.

(5) Other behavioral mutants now under investigation also
include <u>Fast-1</u>, <u>Jerkers</u>, <u>Slow</u> and many other intermediate and leaky
mutants. <u>Fast-1</u> (Kung, 1971a) has an accelerated forward movement
two to three times the normal speed. Rapid forward movement results
from augmentation of ciliary beat in the normal direction, a state
which is known to be correlated with membrane hyperpolarization.
<u>Jerkers</u> are mutants with an increased rate of spontaneous avoiding
reactions with sudden backward dashes. This seems to be another
Na^+-related defect. <u>Slow</u> mutants move about extremely slowly and
often stop. Tamm (personal communication) found that in some
specimens of Slow only the oral cilia seem to be beating but not
the body cilia. This may indicate that there is differential control of the oral and somatic ciliature and that this control also
can be attacked by mutations.

Thus, we have started to build a "library" of mutants with
various defects on a piece of membrane. Our collection, though
small with respect to our goal, already indicates that the project
is feasible. Through physiological studies we aim to identify the
specific molecular lesions in each one of the mutants. In theory,
all membrane proteins functioning in various mechanisms, such as
ion gating, exchange or electrogenic pumping, chemoreception or
other sensory mechanisms, can be so identified. Conditional
mutants, such as temperature sensitive mutants, can be used to circumvent lethality for those membrane functions vital to the survival of the cells. The last part of the project, the isolation
of the gene products has now begun. The identification and characterization of the molecular component responsible for the voltage
sensitive gating mechanisms defective in Pawns, for instance, will
be of great interest to all neurobiologists. For review of this
work see Kung (1974) and Kung <u>et al</u>. (1974).

SUMMARY

The genetic approach to neurobiology is based firmly on our
knowledge of molecular biology. Careful examination of the phenotypes
caused by mutational alterations of structural elements can help us
understand how the elements function in a biological system. Although
they have no nervous system, protozoa and bacteria show well defined
behavior and have many of the basic cellular properties of interest
to neurobiologists. These unicells are eminently suitable for
genetic analysis. Work on <u>Paramecium</u> <u>aurelia</u> and <u>Escherichia</u> <u>coli</u>

has shown how genetics can be used as a powerful tool in the search
for molecular mechanisms in neurobiology.

REFERENCES

Adler, J. 1969. Chemoreceptors in bacteria. _Science_, _166_, 1588–
  1597.
Adler, J., Hazelbauer, G.L. and Dahl, M.M. 1973. Chemotaxis toward
  sugars in _Escherichia coli_. _J. Bact._, _115_, 824–847.
Armstrong, J.B. and Adler, J. 1967. Genetics of motility in
  Escherichia coli: complementations of paralyzed mutants. _Genetics_,
  _56_, 363–373.
Armstrong, J.B. and Adler, J. 1969. Complementation of nonchemotactic mutants of _Escherichia coli_. _Genetics_, _61_, 61–66.
Armstrong, J.B., Adler, J. and Dahl, M.M. 1967. Nonchemotactic
  mutants of _Escherichia coli_. _J. Bact._, _93_, 390–398.
Benzer, S. 1967. Behavioral mutants of _Drosophila_ isolated by
  counter-current distribution. _Proc. Natl. Acad. Sci. U.S._, _58_
  1112–1119.
Brenner, S. 1973. The genetics of behavior. _Brit. Med. Bull._, _29_,
  269–311.
Chang, S.Y. and Kung, C. 1973a. Temperature sensitive Pawns: heat-
  sensitive behavioral mutants of _Paramecium aurelia_. _Science_, _180_,
  1197–1199.
Chang, S.Y. and Kung, C. 1973b. Genetic analyses of heat sensitive
  pawn mutants of _Paramecium aurelia_. _Genetics_, _75_, 49–59.
Eckert, R. 1972. Bioelectric control of ciliary activity. _Science_,
  _176_, 473–481.
Eckert, R., Naitoh, Y. and Friedman, K. 1972. Sensory mechanisms
  in _Paramecium_ I. Two components of the electric response to
  mechanical stimulation of the anterior surface. _J. Exp. Biol._,
  _56_, 683–694.
Hazelbauer, G.L. and Adler, J. 1971. Role of galactose binding
  protein in chemotaxis of _Escherichia coli_ toward galactose.
  Nature New Biology, _230_, 101–104.
Jennings, H.S. 1906. _Behavior of the lower organisms_. Columbia
  University Press, New York.
Kamada, T. 1934. Some observations on potential differences across
  the ectoplasm membrane of _Paramecium_. _J. Exp. Biol._, _11_, 94–102.
Kinosita, H., Dryl, S. and Naitoh, Y. 1964a. Changes in membrane
  potential and the responses to stimuli in _Paramecium_. _J. Fac. Sci._,
  Univ. Tokyo, Sect., IV, _10_, 291–301.
Kinosita, H., Dryl, S. and Naitoh, Y. 1964b. Relation between the
  magnitude of membrane potential and ciliary activity in _Paramecium_.
  J. Fac. Sci., Univ. Tokyo, Sect. IV, _10_, 303–309.
Kinosita, H. and Murakami, A. 1965. Control of ciliary motion.
  Physiol. Rev., _47_, 53–82.

Kung, C. 1971a. Genic mutants with altered system of excitation in *Paramecium aurelia* I. Phenotypes of the behavioral mutants. *Z. vergl. Physiologie*, *71*, 142-164.

Kung, C. 1971b. Genic mutants with altered system of excitation in *Paramecium aurelia* II. Mutagenesis, Screening and Genetic analysis of the mutants. *Genetics*, *69*, 29-45.

Kung, C. 1974. Genetic dissection of the excitable membrane of *Paramecium*. *Genetics* suppl. (in press).

Kung, C., Chang, S.Y., Satow, Y., Vanhouten, J. and Hansma, H. 1974. *Science* (in preparation).

Kung, C. and Eckert, R. 1972. Genetic modification of electric properties in an excitable membrane. *Proc. Natl. Acad. Sci.*, *69*, 93-97.

Kung, C. and Naitoh, Y. 1973. Calcium-induced ciliary reversal in the extracted models of "Pawns" a behavioral mutant of *Paramecium*. *Science*, *179*, 195-196.

MacNab, R.M. and Koshland, D.E., Jr. 1972. The gradient-sensing mechanism in bacterial chemotaxis. *Proc. Natl. Acad. Sci.*, *69*, 2509-2512.

Mesibov, R. and Adler, J. 1972. Chemotaxis toward amino-acids in *Escherichia coli*. *J. Bact.*, *112*, 315-326.

Naitoh, Y. and Eckert, R. 1968a. Electrical properties of *Paramecium caudatum*. Modification by bound and free cations. *Z. vergl. Physiol.*, *61*, 427-462.

Naitoh, Y. and Eckert, R. 1968b. Electrical properties of *Paramecium caudatum*: All-or-none electrogenesis. *Z. vergl. Physiol.*, *61*, 453-472.

Naitoh, Y. and Eckert, R. 1969a. Ionic mechanism controlling behavioral responses of *Paramecium* to mechanical stimulation. *Science*, *164*, 963-965.

Naitoh, Y. and Eckert, R. 1969b. Ciliary orientation: controlled by cell membrane or by intercellular fibrils. *Science*, *166*, 1633-1635.

Okajima, A. and Kinosita, H. 1966. Ciliary activity and coordination in *Euplotes eurystomus* I. Effect of microdissection of neuromotor fibres. *Comp. Biochem. Physiol.*, *19*, 115-131.

Satow, Y., Chang, S.Y. and Kung, C. 1974. Membrane excitability: made temperature dependent by mutations. *Proc. Natl. Acad. Sci. U.S.* (in press).

Satow, Y. and Kung, C. 1974. Genetic dissection of active electrogenesis in *Paramecium aurelia*. *Nature*, *247*, 69-71.

Sonneborn, T.M. 1970. Methods in *Paramecium* research. In: *Methods of Cell Physiology*. Vol. 4. D.M. Prescott (Ed.), Academic Press, New York.

Taylor, C.V. 1920. Demonstration of the function of the neuromotor apparatus in *Euplotes* by the method of microdissection. *Univ. Calif. Publs. Zool.*, *19*, 403-471.

Tso, W.W. and Adler, J. 1974. Negative chemotaxis in *Escherichia coli*. *J. Bact.*, *118*, 560-576.

CYBERNETICS AND THE BEHAVIOR

OF MICROORGANISMS

Bodo Diehn
Department of Chemistry
The University of Toledo
Toledo, Ohio 43606

INTRODUCTION

The most obvious advantage of studying sensory phenomena in unicellular rather than in higher organisms is, of course, the fact that the processes involved in sensory transduction are much less complex, and thus presumably more easily understood, in the former. When I first became interested in the subject, it was with the now common idea that the stimulus/response systems of unicells might be suitable models for the sensory systems of higher organisms. While this view is justified to a significant extent (as will be evident to the reader of this book), I soon became thoroughly fascinated as well with the unique mechanisms which motile unicells have evolved to utilize information from their environment for seeking out favorable external conditions. Moreover, the analysis of these sensory systems has, for one or two organisms, now reached the stage where we have been able to devise computer models and cybernetic analogues of the organisms which exhibit all of their known properties with respect to motor responses toward stimulation. While this goal is as yet unattainable in the case of higher organisms, the approaches described below could well prove useful for future attempts in that direction.

THE MOTOR RESPONSES OF STIMULATED UNICELLS

The movement of microorganisms which results in accumulation in or dispersal from stimulated regions need not in itself be directed. In fact, truly directed movement, known as "taxis", has only been demonstrated in eucaryotes. Procaryotic organisms

apparently have not evolved the more complex control mechanisms
that are required for oriented movement.

Since I wish to report on studies of the receptor/effector
chains that control motile behavior, I shall start with a short
description of possible behavioral responses of unicells. The
classification which follows is a modification of the scheme
first proposed by Fraenkel and Gunn (1940):

1. Nondirectional Responses

<u>Klinokinesis</u>. This is the modulation by the stimulus of the
rate of directional change (R.D.C.) of randomly moving organisms.
The phenomenon is known as "biased random walk" in physics. In
"inverse" klinokinesis, the R.D.C. decreases upon an increase in
stimulus intensity. As a consequence, the cells change direction
less often when moving up a stimulus gradient, and thus eventually
accumulate in the region of highest stimulus intensity. This type
of behavior is, for instance, responsible for chemoaccumulation of
<u>E. coli</u> (Berg and Brown, 1973). In "direct" klinokinesis, the
R.D.C. increases with increasing stimulus intensity, thus leading
to dispersal from a stimulated region.

<u>Kinesis</u>. In this response, the stimulus modulates the linear
velocity of movement. Here again, an inverse kinesis leads to
accumulation in the stimulated region because the organisms slow
down and thus experience a delay in leaving this zone. Analogously,
a direct kinesis leads to a dispersal from the stimulated region,
as in photostimulation of <u>Rhodospirillium</u> (Throm, 1968).

<u>Phobic Responses</u>. They are shock reactions which occur as the
consequence of changes in stimulus intensity. They are expressed
as cessation of forward movement, followed by turning and subsequent movement in another but random direction. Phobic responses
are distinct from klinokinesis in that the response immediately
follows the change in stimulus intensity, occurs only once, and is
exhibited only in certain ranges of stimulus intensity. An inverse
phobic response is one which is evoked by a decrease in stimulus
intensity. It leads to accumulation, since a cell which encounters
a region of lower stimulus intensity is essentially reflected back
into the stimulated region. No such barrier exists when the same
cell experiences the increase of stimulus intensity upon moving in
the opposite direction, since direct and inverse phobic responses
are never exhibited in the same range of stimulus intensity. The
latter fact indicates that the expression of phobic responses
requires sensory systems which are sufficiently complex to discrim-
inate upper and lower thresholds. Phobic reactions are, in fact,
usually exhibited only by eucaryotic cells. I will discuss one such
system, <u>Euglena</u>, in detail.

2. Directional Responses

Responses which involve directional homeostasis, i.e., orientation and subsequent oriented movement with respect to the direction of the stimulus, are called "Taxes". A true taxis never has been observed in procaryotes, whose sensory transduction apparatus apparently lacks the sophistication which is required for maintaining alignment. In this context, the term "chemotaxis of bacteria" is a misnomer, since the phenomenon of chemoaccumulation is based on klinokinesis.

In the following, I will discuss in some detail the analysis of sensory transduction phenomena in _Salmonella_ and in our own pet bug, _Euglena_.

SYSTEMS ANALYSIS OF SENSORY TRANSDUCTION IN ANEURAL SYSTEMS

The stimulus/response system of higher organisms is commonly divided into three component parts: The input (receptor), the internuncial part (nervous system), and the output (muscle). While quite a bit of information has been collected about the receptor and output functions in microorganisms, very little is known about the possible existence in these cells of an analogue to the internuncial part. However, one can determine, from studies of input/output relationships, whether anything occurs in the utilization of sensory information by the cell which would be indicative of signal processing. At the present, only our work with the flagellated photosynthetic alga _Euglena_ has been specifically directed toward this goal. No systematic studies of behavioral responses as a function of variation in stimulus intensity etc., have been conducted with other unicells. However, Macnab and Koshland (1972) recently have published some data on a bacterial system which permit a few preliminary working hypotheses as outlined below.

Klinokinesis in _Salmonella Typhimurium_

Chemoaccumulation toward serine of this procaryotic organism is mediated by inverse klinokinesis. When the attractant concentration is rapidly (within 200 msec) changed from 1.0 to 0.24 mM in a stopped-flow type apparatus, the Rate of Directional Change (R.D.C.) increases from the basal value to a higher, but not quantitatively measured level, and decays within about 12 seconds back to the basal rate.

If the attractant concentration is increased from 0 to 0.76 mM, the R.D.C. is depressed significantly, but returns to the basal value within about 300 seconds. Linear velocity is unaffected by stimulus intensity.

What do these data tell us about the sensory mechanism in
<u>Salmonella</u>?

First of all, the receptor-effector chain appears to contain a
"Directional Change Generator" which, when running free, acts on
an effector such as to vary flagellar (i.e., motor) coordination
randomly but at a constant rate. The D.C.G. function is not carried
out by a specific organelle though it might be localized in the
membrane as are, e.g., the chemoreceptors of <u>E</u>. <u>coli</u> (Kalckar, 1971).
Moreover, modulation of the D.C.G. occurs via an as yet unidentified
device which I will call the "processor" and which is capable of
sensing temporal changes in stimulus intensity as transmitted by
the receptor (Figure 1). To do so the processor has to compare
the receptor signals which it senses at different times -- a process
which requires the storage of information for a certain length of
time. Such storage can, in principle, be accomplished by transducing the receptor signal into any physical or chemical process
which has a slow relaxation time. Simultaneous transduction into
a second and fast process of the same nature as the slow component
allows the requisite temporal comparison and generation of a
physical or chemical difference signal which serves to modulate the
D.C.G. Macnab and Koshland have proposed a model based on conformational changes induced by the stimulus in enzymes which control
the pool size of a compound that effects flagellar coordination.
In Figure 2, a modification of their scheme is shown which can
account for the different relaxation times of positive versus
negative modulation.

The above scheme is quite simple, yet it contains at least
6 kinetic parameters of two allosteric enzymes. As more quantitative
data become available, modifications of the model will undoubtedly
become necessary. If one wishes to quantitatively test such schemes,
it is advantageous to treat them as if the processor were an electronic device. This simplifies the model and makes it amenable to
computer simulation, with the concomitant advantages of a rapid
fit to new experimental data.

As an example, let us assume that the receptor signal is
electrical in nature, and that the intensity information is stored
by changing a storage device, e.g., a membrane capacitance. The
latter would be the slow process, while the immediate appearance
of the receptor signal proper represents the fast component.

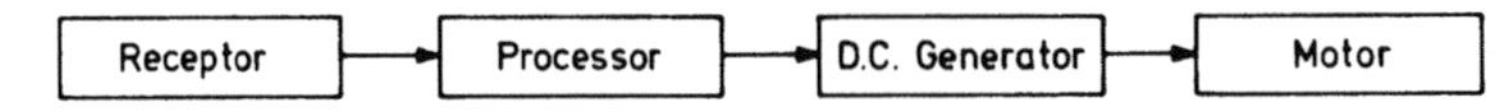

Figure 1. Components of the sensory system of <u>Salmonella</u>.

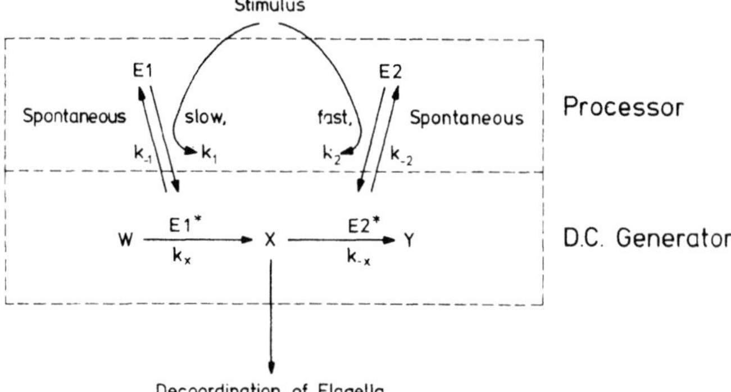

Figure 2. A proposed scheme for flagellar control in Salmonella based on temporal comparison of stimulus intensities (after Macnab and Koshland, 1972, modified). Stimulation converts enyzmes 1 and 2, at different rates, to catalytically more active forms. Since E2 is converted faster, this results in a decrease in the concentration of compound X which is responsible for the loss of flagellar coordination. The relation between k_1 and k_2 is such that the steady state value of X is reattained faster after stimulation than after removal of the stimulus, where the transient increase of [X] is controlled by k_{-1}/k_{-2}.

Modulation of the D.C.G. which is proportional to the direction and magnitude of the charge rate on the capacitance completes the processes occurring in our electronic analogue of the sensory system.

This analogue is shown as a flow sheet in Figure 3. Note that only three yes/no decisions are required in computer simulation. For example, the first decision, "Is the stimulus intensity increasing or decreasing?" would be made by determining the sign of the difference between receptor and storage potentials.

Hypothesizing an electrical nature of stimulus transduction in unicells is not totally unreasonable, as Dr. Wood's chapter in this book demonstrates. However, once a functioning computer model has been devised, it can always be translated back into systems of interrelated biochemical reactions, if doing so should facilitate the design of further experimental tests of the theory. The latter is, of course, the main reason for conducting simulations at all.

In the case of Salmonella, insufficient data in the literature and of course the fact that no work with bacteria is being done in our laboratories, are responsible for the fact that we have not been conducting actual simulations. However, bacteria will be very important and attractive systems for studying sensory transduction, particularly because of the sophisticated genetic analysis which is possible with organisms such as E. coli.

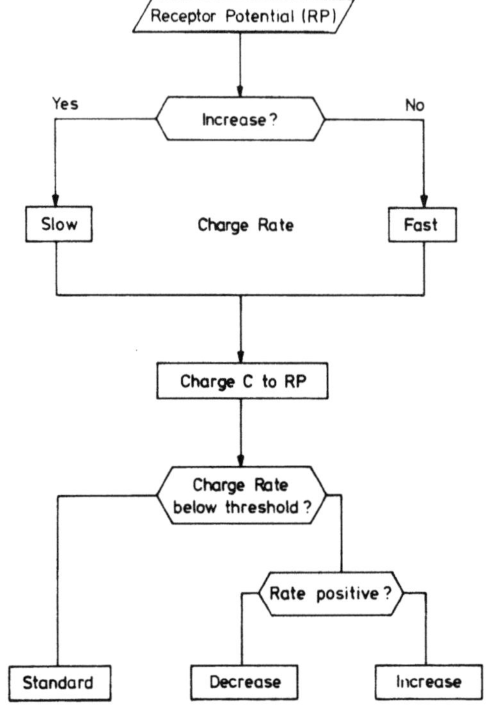

Figure 3. An electronic analogue of the scheme in Figure 2 (see
text).

Photophobic Responses of _Euglena_ _Gracilis_

The major disadvantage of chemical stimulation is that it
cannot be varied and sequenced at will, thus making difficult a
systematic study of stimulus/response relationships. Light is a
stimulus which is not only manipulated with much greater ease, but
which can also be removed completely at any desired instant. For
these reasons, work in our laboratories is mainly concerned with
photoresponses. The experimental organism for most of our studies
has been the photosynthetic flagellate _Euglena_ (Figure 4). This
eucaryotic cell is capable of true taxis, i.e., it can display
orientation and subsequent oriented movement with respect to the
direction of the light source. Since in the context of this
chapter I should restrict my discussion to interrelationships
between the components of the sensory system of our organism, I
must refer you to a chapter in another book (Diehn, 1972) for a
summary of our work on the biochemistry and physiology of photoreception and related phenomena. It must suffice here to say that

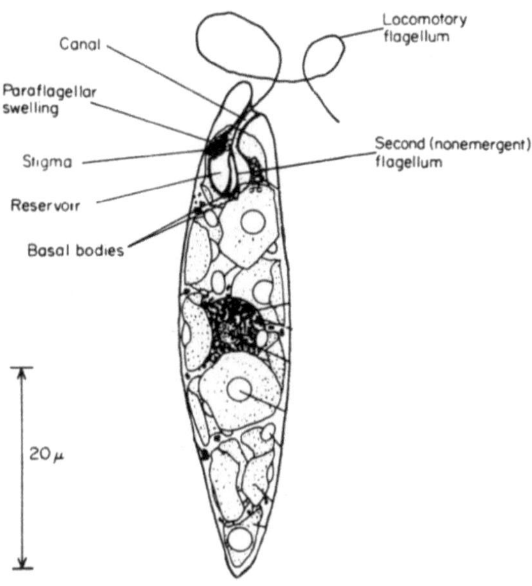

Figure 4. Morphology of _Euglena_ _gracilis_ (after Leedale).

the photoreceptor function is apparently carried out by the
paraflagellar swelling, which is enclosed within the flagellar
membrane and situated at the junction of the two flagella.

Because of the asymmetric thrust of the flagellum, the cell
rotates around its longitudinal axis during forward motion. The
photophobic reaction of _Euglena_ is expressed as a cessation of
forward movement, followed by rapid turning around a lateral axis
toward the "dorsal" side which bears the stigma.

After an adaptation time which depends upon the magnitude and
direction of the illumination change and on the excitation state of
the cell's sensory system (see below), the turning stops and
forward movement is resumed in a new and random direction. Photophobic responses are, besides through adaptation, also terminated
by restoration of the previous lighting conditions. Positive
Phototaxis involves the following sequence of events: If the cell
is stimulated laterally with light of moderate intensity, the shadow
of the stigma will fall upon the paraflagellar swelling once every
revolution. This will result in a decrease of the incident light
intensity, and thus cause an inverse photophobic response which
will cease when the photoreceptor is no longer shaded. The result
is a fractional turn toward the dorsal side, i.e., toward the side
from which the light came when shading occurred. Repetition of this
process results in further course corrections which end when the cell

is oriented directly toward the light source, at which time
shading of the photoreceptor by the stigma has become geometrically
impossible.

Once the cell is oriented toward the stimulus, any deviation
from the proper direction results in a shading event. This generates
an error signal that is transmitted to the motor apparatus and
triggers the course correction which is required for directional
homeostasis (Diehn, 1969).

It is evident that in order to elucidate the transduction
properties of the light sensing system of _Euglena_, one should study
the photophobic responses as a function of stimulation. As a preliminary step, one must define the minimum components of the
sensory transduction system that mediates photomovement of _Euglena_,
as well as the probable interactions between these components.

The sensory chain starts with the photoreceptor, whose output
is modulated by the shading action of the stigma. The photoreceptor signal is then converted in the processor to a command
which acts on an effector such as to cause reorientation of the
flagellum. A source of energy for the system also will have to be
incorporated in the scheme. This requirement arises from the
results of our studies of the effects on photo-accumulation of
substances that affect biological energy transduction. In this
work we demonstrated that flagellar motion requires the products
of oxidative phosphorylation, while a functioning photosystem II
of photosynthesis, more specifically non-cyclic photophosphorylation, is required for photo-accumulation and hence, for the inverse
photophobic response to occur (Diehn and Tollin, 1967). Non-
photosynthetic mutants are incapable of inverse responses, but can
still exhibit direct photophobic reactions.

In the resulting scheme, which is shown in Figure 5, I propose
as a working hypothesis that an "inverse effector" is energized by
a product of photosynthesis, while the "direct effector" depends
on respiration for its supply of energy. At this point it is not
really necessary to postulate two effectors; the site of interaction
between energy production and sensory transduction could be moved
to the processor without significant alteration of the overall
scheme. The feedback loop between motor and modulator indicates
that the motor apparatus is responsible for changes in the relative
positions of modulator and receptor with respect to the light
source.

As pointed out before, the combination of modulator and receptor generates an error signal if the cell deviates from the
proper orientation, and the motor apparatus responds in such a way
as to minimize this error signal. While the above homeostatic

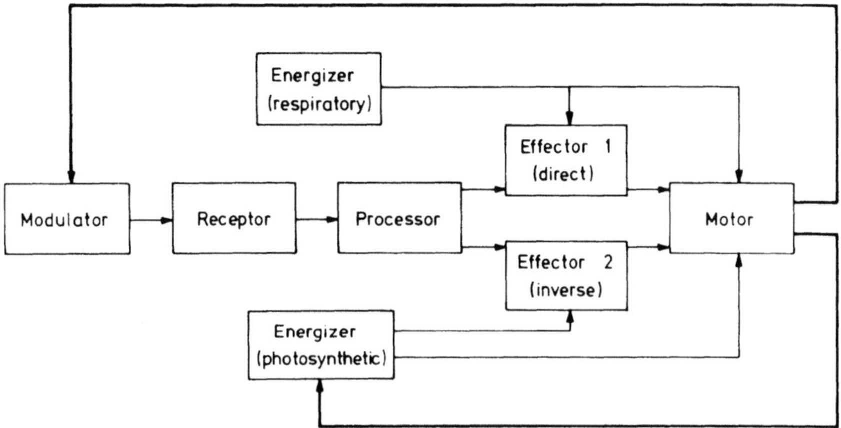

Figure 5. Flow diagram of interactions within the sensory transduction chain involved in photomovement of Euglena.

system is characterized by the flow of information between its
components, there exists another feedback loop, between the motor
and the photosynthetic apparatus, which is characterized by the
flow of energy: When Euglena is illuminated with white light of
moderate intensity, the linear swimming velocity increases within
about 15 minutes (Wolken and Shin, 1958). This type of photokinesis can be explained by movement to the flagellum of high-
energy compounds produced in photosynthesis.

The scheme shown in Figure 5 contains two components that have
not yet been identified in the cell: The effector(s) and the processor. It seems reasonable to assume that the effectors will be
identified through light or electron micrographic studies, since
in order to bring about reorientation of the locomotory flagellum
they must be capable of movement and should exhibit the characteristic structure of motile elements. One would expect to find the
effector elements either within the flagella or at their bases.

Information about the properties of the signal processor can
be derived from studies in which the photophobic responses of the
cells are observed during sequential stimulation. Table 1 summar-
izes the results of such a series of experiments which were conducted during our investigations into the mechanism of phototactic
orientation (Diehn, 1969). An analysis of the data of Table 1
reveals the following features:

1) A single light pulse of low intensity will reset the
transduction system for the inverse phobic reaction without in-
ducing a direct photophobic response. Conversely, a "dark pulse"
during high-intensity illumination resets for the direct phobic
response.

TABLE 1

Photophobic responses of _Euglena_ as a function of light intensity
and pre-adaptation.

		Intensity of White Light [erg/cm^{2}sec]		
		8 x 10^5	2 x 10^5	5 x 10^3
Light-Adapted	Light Off	No immediate shock, slight delayed shock	50% shock	Shock
	Light on after 1.0 sec dark	Shock	Stops shock, Shocks previously Un-shocked cells	Stops Shock
Dark-Adapted	Light on	Shock	50% shock	No Shock
	Light Off after 1.0 sec illumination	Stops Shock, Delayed 2nd Shock	Stops Shock, Shocks previously Un-shocked cells	Shock

2) If the system is adapted to any light intensity, a change
in intensity induces a photophobic reaction only when the direction
of change is away from an "adaptation level" which for the particular culture studied was about 2 x 10^5 erg/cm^2 seconds of white
light from a xenon lamp. It appears as though the system were
naturally adapted to this light intensity which not unexpectedly
corresponds to the saturating intensity for photosynthesis, even
though the cells had for many generations been cultivated at much
lower levels of illumination. On the assumption that at the output
of the photoreceptor there appears an electrical signal which is
a function of the incident light intensity (Diehn and Kint, 1970),
the processor must be capable of comparing the magnitude of this
signal with that of an internal reference level and of determining
the sign of $\frac{dI}{dt}$ in order to activate the effector.

3) A system that has adapted to a low light intensity will
exhibit an immediate direct photophobic response upon an increase
of the intensity beyond the adaptation level, while after adaptation to the high intensity darkening will elicit only a delayed
and attenuated inverse reaction.

4) If a direct photophobic response is induced by illumination
with high intensity, darkening after 1 second will not result in
an immediate transition to the inverse response. Conversely, we
found that transition to the direct response is not observed upon
illumination with high intensity of a cell that has just commenced
a strong inverse response.

The operational necessities and advantages for the cell of
such complex behavior are as follows:

1) The intensity ranges in which direct and inverse responses
occur are clearly separated, thus permitting accumulation in or
dispersal from stimulated regions (depending on light intensity)
on the basis of phobic reactions alone.

2) The resetting mechanism allows directed movement away from
a high intensity light source, i.e., _negative_ phototaxis, as
follows: If the intensity of the lateral illumination is above the
adaptation level, the cell will exhibit a direct photophobic response.
Since every shading event stops this response but resets the cell's
sensory mechanism for a further direct response upon re-illumination
of the photoreceptor, the only situation in which there will be no
further phobic response is one in which the photoreceptor is
permanently shaded. This can best be accomplished by the posterior
end of the cell and requires the organism to be oriented directly
away from the light source.

The behavior of the cell not only indicates that signal processing occurs but also yields information on the dynamic characteristics of the processor. An attempt can now be made to use the
techniques of Systems Analysis in order to elucidate the approximate
internal structure of the processor "black box" (Diehn, 1973).

The input of the processor will be taken to consist of an
electrical signal whose magnitude is a function of the incident
light intensity. The output consists of commands which activate
the appropriate effector.

An internal "reference level" in the processor corresponds
to the photoreceptor signal at the adaptation level of light
intensity. A sequence of processing steps which will generate
effector control signals that exactly duplicate the response of
the actual receptor/effector system of _Euglena_ as summarized in
Table 1, is shown in Figure 6.

While this diagram appears complex, it describes in essence
only the charge or discharge of two storage devices (e.g., membrane
capacitances) toward the receptor or reference potential level as
shown. If the charge rate exceeds a threshold value, the effector

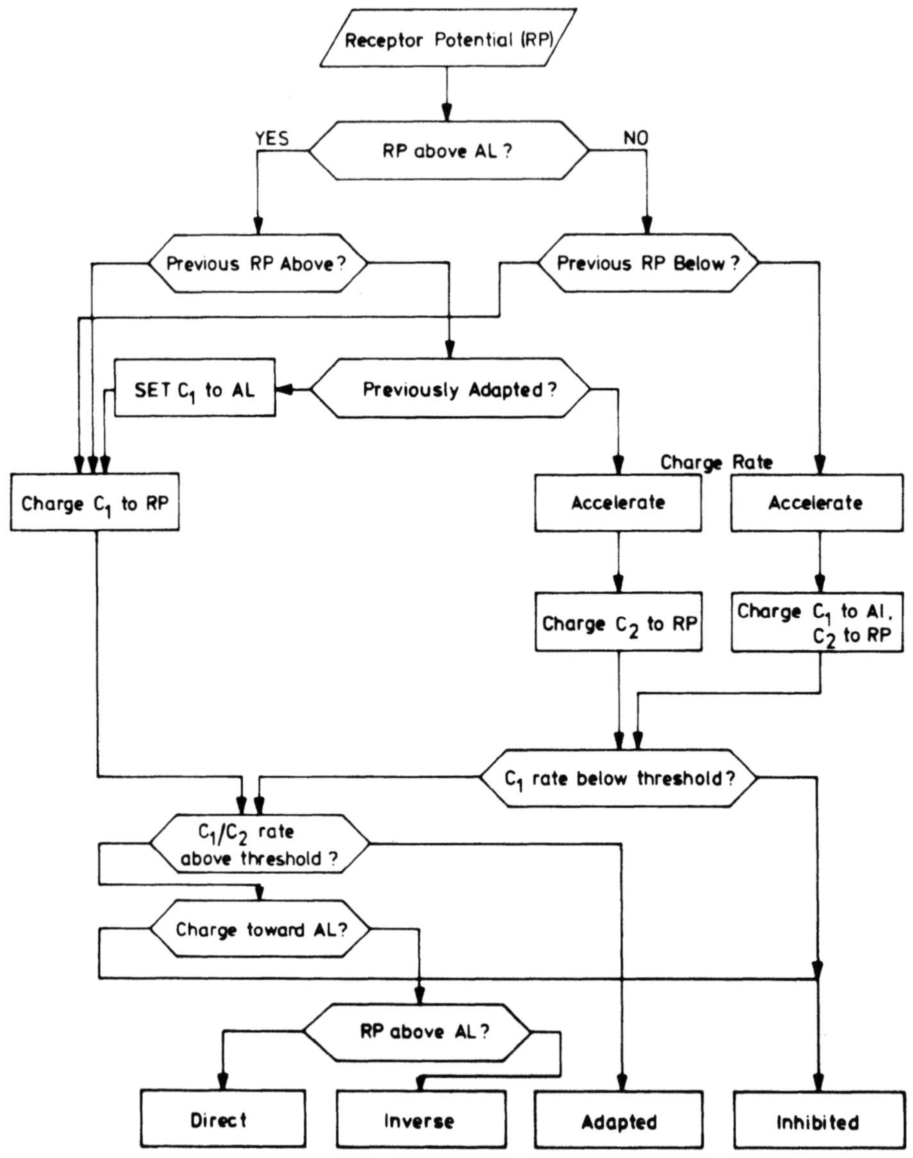

Figure 6. Computer flow sheet of signal processing steps which are presumed to link photoreception and the expression of photophobic responses in _Euglena_. RP, Receptor Potential; AL, Adaptation Level.

is activated; otherwise the system is considered adapted. The second of the capacitances is utilized only when a change in light intensity from below to above the adaptation level, or vice versa, is imposed upon the system while it has not yet adapted to the previous illumination, i.e., when the first capacitance is incompletely charged or discharged. When both storage devices are being utilized, charge and discharge occur at an accelerated rate and the signal activating the effector is suppressed.

While translation of this scheme into one of interrelated chemical reactions would be possible, the latter would be no less hypothetical than the former until more experimental evidence has been accumulated with respect to the nature of the transduction process. We are at present using electrophysiological techniques in an attempt to detect action potentials or analogous phenomena upon stimulation.

The scheme of Figure 3 has been translated into a FOCAL computer program for execution with a PDP-8 digital computer. The printout of a simulation of the "light-adapted/light off" and "dark-adapted/light on" experiments in the "8 x 10^5" column of Table 1 is shown in Table 2. Note that system dynamics are characterized by only four variables. The agreement between observed and simulated behavior is striking.

A model is most useful when it can be utilized for predicting system behavior under conditions that have not yet been investigated experimentally. Such a situation is one in which a high light intensity is lowered to a value still above the adaptation level. Our model predicts that no photophobic reaction will occur (the computer printout is not shown, but tracing the scheme of Figure 6 will verify the statement). When the experiment was done, we indeed observed that cells adapted to 6 x 10^5erg/cm^2 sec, did exhibit a direct photophobic response upon an increase to 8 x 10^5, but showed no reaction when the intensity was lowered from the latter value to 4 x 10^5.

A further prediction of our model is that the responses are additive in the sense that if the stimulus intensity is changed in monotonic steps, the sum of the response durations in the individual steps should be equal to the duration of the response if the same change of stimulus intensity were imposed in one step. Table 3 shows the results of an experiment designed to test this prediction. The agreement is acceptable. It is to be noted, however, that the dependence of response duration on intensity change is not exponential as would be expected from a simple capacitor charging model. There appears to be a linear relationship instead. In terms of our model, this means that a constant-current device should be incorporated in the charging circuits. It is interesting to note in this context that the adaptation level (threshold of the direct response) of this particular culture, which had been suspended in an inorganic "resting medium" for three days, had a value of 1.0 x 10^5erg/cm^2 sec. When cells were taken from the same culture and tested 6 hours after

TABLE 2

Computer simulation of "Light adapted/light off" and "Dark adapted/
light on" experiments ("8x10^5" column of Table 1).

PHOTOPHOBIC RESPONSES OF <u>EUGLENA</u>

ADAPTATION LEVEL (LOG I):5.3
CROSS INHIBITION TIME CONSTANT (SEC):.5
EFFECTOR THRESHOLD:.Ø5
EFFECTOR TIME CONSTANT (SEC):2

LOG STARTING ILLUMINATION:5.9
DURATION (SEC):3
ILLUMINATION CHANGE TO (LOG):2
DURATION (SEC):3
LOG FINAL ILLUMINATION:5.9
DURATION (SEC):3

PLOT FROM (SEC):.5
TO (SEC):9

SEC	LOG I	RESPONSE
Ø.Ø	5.3Ø	ADAPTED
Ø.5	5.9Ø	DIRECT
1.Ø		DIRECT
1.5		DIRECT
2.Ø		DIRECT
2.5		ADAPTED
3.Ø		ADAPTED
3.5	2.ØØ	INHIBITED
4.Ø		INHIBITED
4.5		INHIBITED
5.Ø		INVERSE
5.5		ADAPTED
6.Ø		ADAPTED
6.5	5.9Ø	DIRECT
7.Ø		DIRECT
7.5		DIRECT
8.Ø		DIRECT
8.5		ADAPTED
9.Ø		ADAPTED

TABLE 3

Duration of direct photophobic responses of <u>Euglena gracilis</u> as a
function of the change in light intensity.

INTENSITY CHANGE $(erg/cm^2$ sec; white light)		RESPONSE DURATION
From	To	(seconds)
0	2×10^6	40 ± 5
0	3×10^5	6 ± 3
3×10^5	2×10^6	38 ± 5
0	5×10^5	10 ± 2
5×10^5	2×10^6	27 ± 4
0	10^6	19 ± 3
10^6	2×10^6	19 ± 4

resuspension in resting medium, this threshold intensity remained
unchanged, while the response durations were found to be longer by
a factor of 2.

Our model also postulates that a reduction of the light intensity during a direct photophobic response will terminate the response
only if the storage device is already charged to a potential which
matches or exceeds that of the receptor output at the new intensity.
This prediction was investigated experimentally by reducing the light
intensity from 2×10^6 to $5 \times 10^5 erg/cm^2$ sec at various times after
the onset of high intensity illumination. If the intensity reduction was imposed 2 seconds after the onset of stimulation with
$2 \times 10^6 erg/cm^2$ sec, the response continued, while the same reduction
of intensity after 15 seconds immediately terminated the response.

Information on the transduction linearity of the receptor/effec-
tor system also can be obtained from threshold measurements of light
pulses that "reset" for the inverse phobic response. As shown in
Table 4, in the range of 10^2 to $10^4 erg/cm^2$ sec there is an approximately linear relationship between intensity and duration of
pulses of monochromatic light which just cause the reappearance of
an inverse response. The only requirement for resetting appears to
be that the pulse contain approximately 80 erg/cm^2. This corresponds
to a reset threshold of about 6×10^4 photons impinging upon the
photo-receptor area.

TABLE 4

Energy content of pulses of monochromatic light that just reset the
stimulus transduction system of _Euglena gracilis_ for the inverse
photophobic response.

INTENSITY (erg/cm^2 sec, 475nm)	RESET THRESHOLD DURATION (sec)	ENERGY CONTENT OF PULSE (erg/cm^2)
6.0×10^3	0.015	90
1.3×10^3	0.05	75
8.0×10^2	0.10	80
2.5×10^2	0.25	63
1.3×10^2	0.50	75
$5.0 \times 10'$	0.55	28
$3.1 \times 10'$	1.0	31
$1.5 \times 10'$	3.5	52

It is apparent that we do have a working model of the signal
processing system controlling the light-induced motor responses of
Euglena. While this model is at present not described in terms
that allow identification of all physiological processes that are
involved in signal transduction, we will know what to look
for as our experimental techniques become more sophisticated.

The prime questions which we will have to investigate in
future studies are whether signal processing in _Euglena_ is based
on electrical or on chemical phenomena, and whether the properties
of the sensory system are modified as a consequence of repeated
stimulation. If phenomena analogous to learning and memory in
higher organisms can be observed, we will have the advantage of
studying them in a system whose properties may soon be known well
enough for the affected parameters to be identified in molecular
terms.

REFERENCES

Berg, H.C. and Brown, D.A. 1972. Chemotaxis in _E. coli_. analyzed
  by three-dimensional tracking. _Nature_, _229_, 500-504.

Diehn, B. 1969. Phototactic response of _Euglena_ to single and
  repetitive pulses of actinic light: orientation time and
  mechanism. _Exp_. _Cell_ _Research_, _56_, 375-381.
Diehn, B. 1972. Phototaxis in _Euglena_. I. Physiological basis
  of photoreception and tactic orientation. In, _The_ _Behavior_ _of_
  Microorganisms, J. Adler (Ed.), Plenum press.
Diehn, B. 1973. Phototaxis and Sensory Transduction in _Euglena_.
  Science, _181_, 1009-1015.
Diehn, B. and Kint, B. 1970. The flavin nature of the photo-
  receptor molecule for phototaxis in _Euglena_. _Physiol_. _Chem_. _and_
  Physics, _2_, 483-488.
Diehn, B. and Tollin, G. 1967. Phototaxis in _Euglena_. IV. Effect
  of inhibitors of oxidative and photophosphorylation on the rate
  of phototaxis. _Arch_. _Biochem_. _Biophys_., _121_, 169-177.
Fraenkel, G.S. and Gunn, Donald L. 1940. _The_ _Orientation_ _of_
  Animals. Oxford University Press.
Kalckar, H.M. 1971. The periplasmic galactose binding protein
  of _Escherichia_ _coli_. _Science_, _174_, 557-565.
Macnab, R.M. and Koshland, D.E. 1972. The gradient sensing
  mechanism in bacterial chemotaxis. _Proc_. _Natl_. _Acad_. _Sci_. _U.S.A._,
  _69_, 2509-2512.
Throm, G. 1968. Untersuchungen zum Reaktionsmechanismus von
  Phototaxis und Kinesis an _Rhodospirillium_ _rubrum_. _Arch_.
  Protistenk., _110_, 313-371.
Wolken, J.J. and Shin, E. 1958. Photomotion in _Euglena_ _gracilis_.
  I. Photokinesis. II. Phototaxis. _J_. _Protozool._, _5_, 39-46.

CONTROL OF CILIARY ACTIVITY IN

ANEURAL ORGANISMS

Miles Epstein
Department of Zoology
University of Minnesota
Minneapolis, Minnesota 55455

Aneural organisms are subjected to many chemical and mechanical
stimuli; these stimuli give rise to receptor potentials as Dr. Wood
has shown. (See paper by D. Wood in this volume -- Ed.). Changes
in membrane potential in turn lead to alterations in ciliary activity, which is one means of locomotion in many aneural organisms. I
wish to discuss some of the mechanisms by which this ciliary activity is regulated.

The major variables in ciliary activity are the orientation of
the effective stroke and the beating frequency, each of which we will
consider separately. Ciliary orientation refers to the direction
of the power stroke, which transfers maximum force to the medium and
thus determines the direction the cell swims. The action of the
cilia can be likened to oars of a boat, where the direction the boat
moves is opposite the direction the oars push against the water.
Hence an effective stroke moving from the front to the rear of the
ciliate ("normal beating") propels the cell in a forward direction;
an effective stroke moving from the rear to the front ("ciliary
reversal") results in the cell swimming backwards. This ability to
change orientation is restricted to the cilia of protozoa, of some
coelenterates, and of some tunicate larvae (Kinosita and Murakami,
1967; Galt and Mackie, 1971). Frequency of ciliary beating governs
the velocity the cell moves through the water. Experimentally it
has been possible to divorce ciliary orientation changes from
ciliary beating. Cilia of _Paramecium_ and _Euplotes_ in which beating
has been stopped with $NiCl_2$ shifted after membrane depolarization
to a forward-pointing position, indicative of ciliary reversal
(Naitoh, 1966; Naitoh and Eckert, 1969).

The evidence which I will review here supports the hypothesis
(Eckert, 1972) that the ciliary membrane undergoes voltage-dependent
changes in Ca conductance. As a result of these conductance changes,
Ca influx and thus the intraciliary Ca concentration increases.
The elevated Ca concentration in turn produces a change in ciliary
orientation (i.e., ciliary reversal).

It is of interest for neurobiologists to study control of
ciliary orientation because of its analogy to excitation-response
coupling. As with muscle contraction, hormone and transmitter
release, and bioluminescence, the effector response is coupled to
excitation by Ca. The electrical and mechanical events involved in
ciliary activity in protozoa can be studied in detail. The cells
can be readily visualized and penetrated with several microelectrodes.
Bathing solutions can be easily exchanged and a stable population
of cells is available.

We will first consider ciliary orientation. Experiments first
done by Kinosita and his coworkers demonstrated that ciliary reversal occurred in response to membrane depolarization (Kinosita,
1954; Yamaguchi, 1960). Extensive study of changes in ciliary
orientation required beating cilia. However, it is difficult to
photograph beating cilia and to correlate beating with membrane
potential. These difficulties were recently overcome by Eckert
and Naitoh (1970), Epstein and Eckert (1973), and Machemer and
Eckert (1972). The techniques for the simultaneous stimulation of
the cells and high-speed cinematography of the cilia are shown in
Figure 1. These techniques allowed the concurrent registration of
the membrane potential and the position of the cilia. Individual
cineframes were analyzed and traced to determine the beating orientation and the beating cycles. _Euplotes_, a ciliated protozoan, was
chosen because its compound cilia (cirri) are large enough to photograph easily. This ciliate, traced in Figure 2, has groups of
cirri located at different positions on the body surface; each
cirrus contains between 40-120 cilia with a typical "9 + 2" arrange-
ment of microtubules (Roth, 1956, 1957; Gliddon, 1966). Differences
in direction of beating can be distinguished in the frontal cirri
by the differences in their extreme positions and their shape.
Differences in the direction of beating of the anal cirri were seen
mainly in their shape during the beating cycles. The beating
cycles of the cirri were plotted as the extremes of inclination
angle (α). These points represented the two extreme positions of
the cirrus in the beat cycle. Details of the method of measuring
α are found in Epstein and Eckert (1973).

The ciliary response was different depending on the direction
of membrane potential change. In Figure 3 a suprathreshold pulse
of outward current was applied through a microelectrode and the
resulting changes in membrane potential and ciliary activity are
shown. A depolarization produced a reversal in beating direction

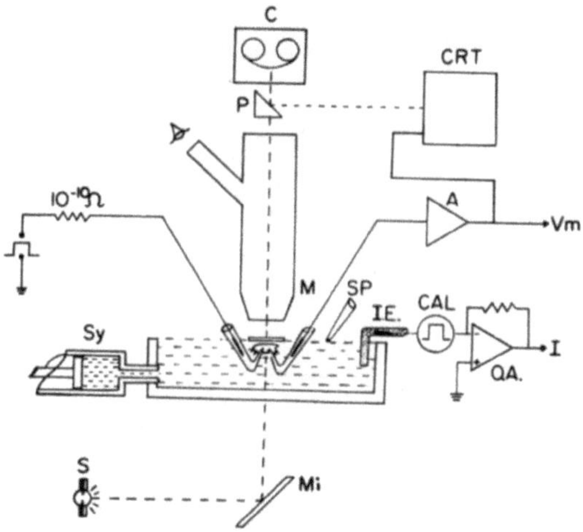

Figure 1. "Diagram of apparatus. The specimen was held against
the lower surface of the coverslip by the voltage-recording and
current-passing microelectrodes. The strobe lamp (S) passed light
through the microscope (M) onto film in the cine camera (C). Membrane potential (Vm) was recorded on a CRT, whose face was projected
onto the cine film through a prism (P). The bath was held at virtual
ground by connecting the indifferent electrode (I.E.) to the summing
junction of an operational amplifier (O.A.) used to monitor current
intensity (I). The cell was bathed in various test solutions injected into the experimental chamber with a syringe (Sy) and removed with a suction pipette (SP). A, neutralized capacity amplifier; CAL, square wave calibrator; Mi, mirror." From Epstein and
Eckert (1973).

of all the compound cilia (this would make an unrestrained cell
swim backward) and an increase in beating frequency. After termin-
ation of the depolarization the cilia returned to the normal beating
direction and the frequency was restored to its original level.

A pulse of hyperpolarizing current resulted in an increase in
beating frequency in all compound cilia while the beating direction
remained normal (so as to make an unrestrained cell swim forward)
as shown in Figure 4. Again the frequency decreased after the
end of stimulus.

Depolarization alone would not produce ciliary reversal; the
presence of Ca in the extracellular solution was also required in
Opalina (Kinosita, 1954), Paramecium (Naitoh, 1968) and Euplotes
(Epstein and Eckert, 1973). This necessity for Ca is illustrated
in Figure 5. If the bath Ca is lowered with EGTA to 10^{-6}M, a

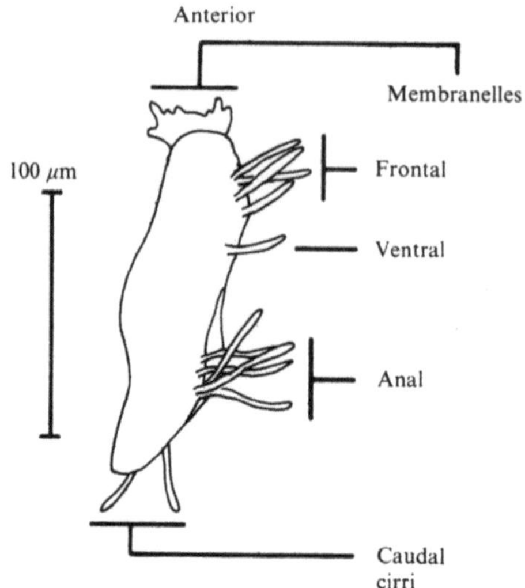

Figure 2. "A tracing from a photomicrograph of _Euplotes_ seen in
side view. The various groups of compound cilia are indicated but
not all cirri of a given group are visible." From Epstein and
Eckert (1973).

Figure 3. "Responses to outward current. A pulse of outward current
(recorded in trace 1 at bottom) resulted in depolarization of the
membrane potential (trace Vm) and changes in the direction and frequency of the beat of the frontal and anal cirri. This activity
in time is plotted as extremes of inclination angle (α) for an anal
cirrus (uppermost plot) and a frontal cirrus (second plot). Inclination angles of the membranelles (third plot) were not measured.
The direction of beating is noted in each case. After the onset of
the stimulus a reversal of the direction and increase in frequency
occurred. The changes in shape of one anal and one frontal cirrus
during a cycle of forward beating before the stimulus are shown in
tracing no. 1 at the left. The numbered bars ($\sqcap$) over the plots of
inclination angles indicate which beating cycles were traced. The
recovery stroke is indicated by the dashed lines. Numbers identi-
fying each position of the cirrus correspond to cine frames counting
from the onset of the current pulse. The frame interval was 5 msec.
The stimulus began at frame 0, and frames before the stimulus are
indicated by numbers 900 to 999. The changes in shape of the same
cirrus during a cycle of reversed beating are shown in tracing no. 2
at the right side of the figure. There the power stroke is directed
toward the front of the cell. Voltage calibration pulse was 20 mV,
50 msec. Current pulse was 1×10^{-9}A." From Epstein and Eckert
(1973).

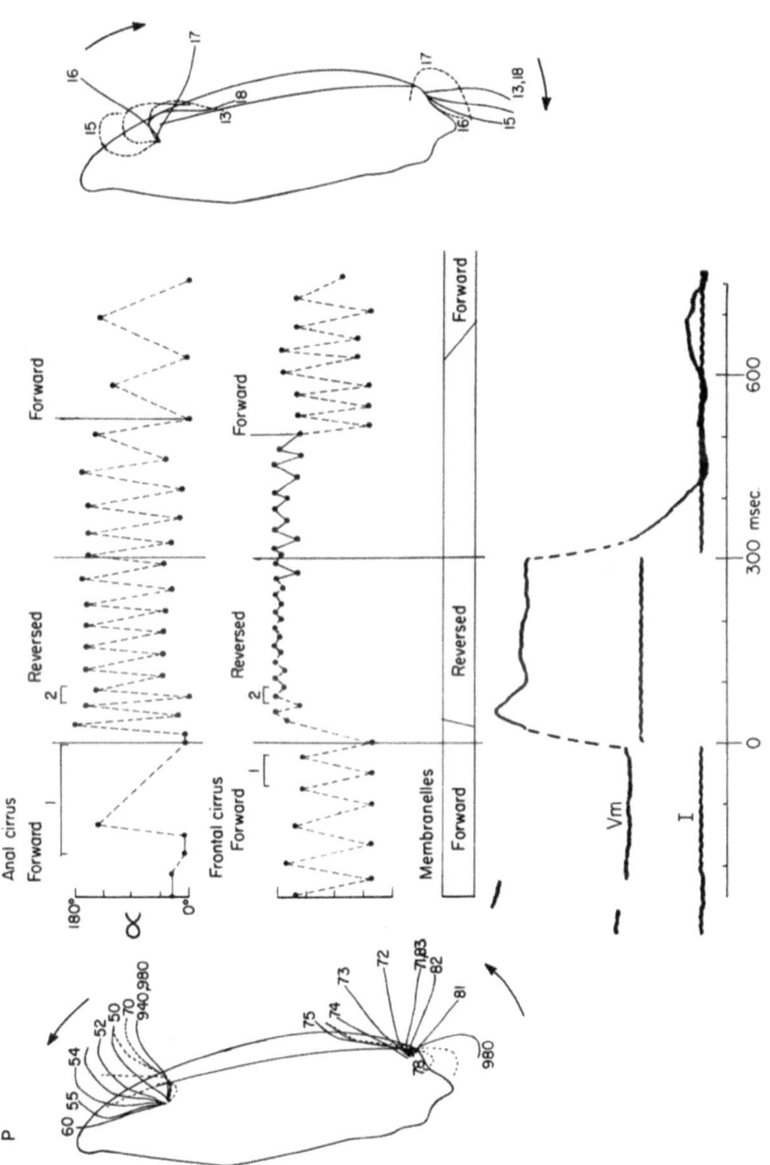

16
17
15
13
18
17
13,18
16
15
Forward
Forward
Reversed
Reversed
Forward
Forward
Anal cirrus
Frontal cirrus
Membranelles
2
2
1
1
180°
0°
α
600
300 msec.
0
Vm
I
P
60 55
54
52
50
70
94 0,980
75
74
73
72
71,83
82
81
78
980

depolarization of the membrane failed to result in ciliary reversal. When the bath Ca is returned to 10^{-3}M, a stimulus resulted in ciliary reversal.

The effect of Ca itself on the ciliary apparatus was examined by Naitoh and Kaneko (1972) in extracted models of <u>Paramecia</u> (Figure 6). The membranes of the models had been disrupted with the detergent Triton X so that ions and metabolites could readily pass into the cell. With addition of ATP and Mg, the cilia were reactivated so that the cells' swimming direction was forward and velocity maximal. As the Ca concentration increased to 10^{-6}M, the velocity decreased and at greater than 10^{-6}M backward swimming was observed. These workers demonstrated in cell models that the beating direction changed from forward to backward as the Ca concentration was raised above 10^{-6}M. Beating frequency, in contrast, did not vary with Ca concentration as shown in Figure 7. Epstein and Eckert (1973) have also found a change in orientation dependent on Ca concentration in reactivated models of <u>Euplotes</u>.

Figure 4. "Responses to inward current. A pulse of inward current (Trace 1) hyperpolarized the membrane potential (trace Vm). The activities of the anal and frontal cirri, expressed as extremes of inclination angle (α), are shown in the first and second plots. Inclination angles of membranelles were not measured. The beating direction of the cirri and membranelles is indicated. The frequency of beating increased during hyperpolarization and slowly decreased after the end of the 300 msec stimulus. A beat cycle of an anal (top) and frontal (bottom) cirrus before (tracing no. 1 at left) and during (tracing no. 2 at right) the stimulus are shown. The numbered bars ($\sqcap$) over the plots of inclination angles indicate which beating cycles were traced. The voltage calibration pulse in Text-figure 3 applies to this figure also. The current pulse was 1×10^{-9}A." From Epstein and Eckert (1973).

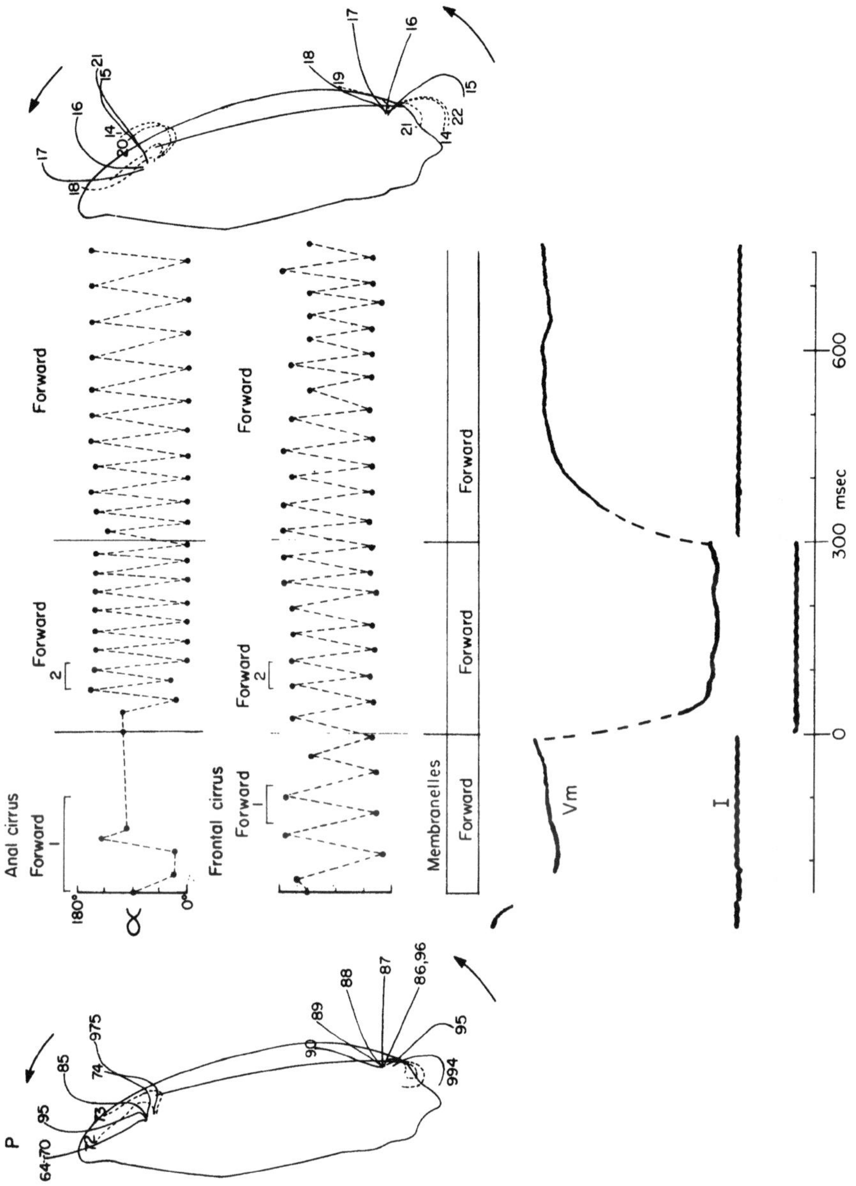
Anal cirrus
Forward
180°
α
0°
Forward
2
Forward
Frontal cirrus
Forward
Forward
2
Forward
Membranelles
Forward
Forward
Forward
Vm
I
0
300 msec
600
P

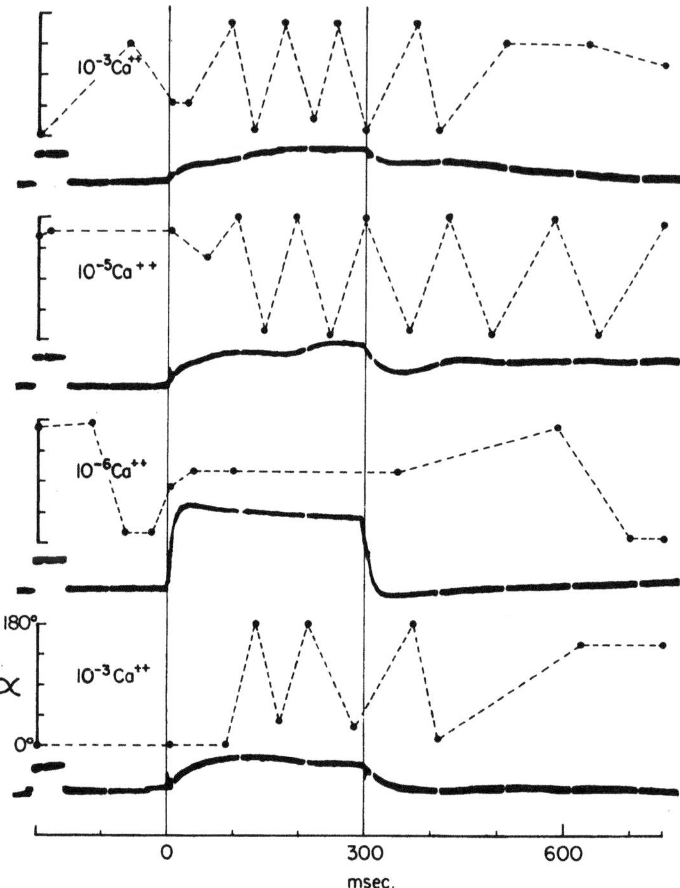

Figure 5. "Loss of depolarization-induced reversed beating response
in reduced external Ca. Membrane depolarizations of a single cell,
bathed in solutions of various Ca concentrations, resulted in ciliary
activity, which is plotted as extremes of inclination angle (α).
When the free Ca concentration was reduced to 10^{-6}M with 10^{-3}M-EGTA,
a 30 mV depolarization did not produce reversed beating. After
returning the Ca concentration to 10^{-3}M, a depolarization of about
10 mV resulted in reversed beating. The resting potential and K
concentration associated with each Ca concentration were -28 mV in
10^{-3}M-Ca, 2 x 10^{-3}M-K; -45 mV in 10^{-5}M-Ca, 2 x 10^{-4}M-K; -35 to -16
mV in 10^{-6}M-Ca + EGTA, 6 x 10^{-5}M-K; and finally -10 mV upon return
to 10^{-3}M-Ca and 2 x 10^{-3}M-K. Voltage calibration was 10 mV, 50 msec.
The intensities of the current pulses were different." From Epstein
and Eckert (1973).

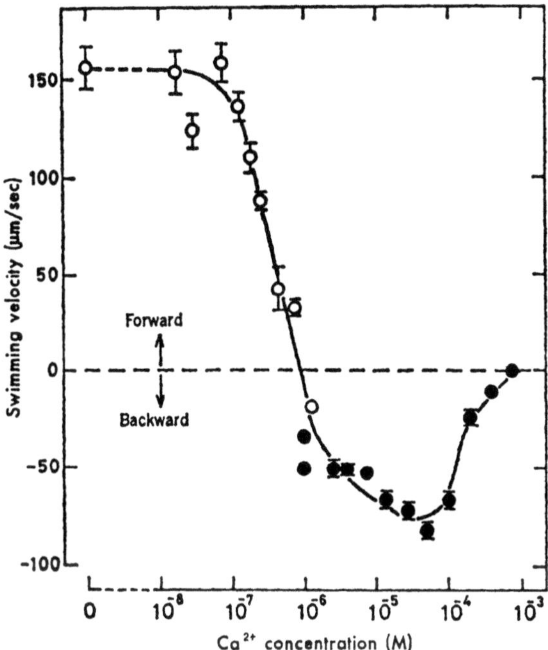

Figure 6. Swimming direction and velocity of reactivated models
of _Paramecium_ are shown as a function of Ca concentration. Ca
was varied with Ca buffers (EGTA) between 10^{-8} to 10^{-6}M (open
circles) and by the simple addition of Ca above 10^{-6}M (solid circles).
From Naitoh and Kaneko (1972).

How then is the intracellular Ca concentration modulated? There
is a large Ca concentration difference directing Ca into the cell;
a change in membrane potential could lead to an increase in Ca
conductance and Ca influx. Naitoh _et al._ (1972) found that stimulation of _Paramecia_ with depolarizing current pulses produced graded
regenerative responses. An illustration of these responses is
given in Figure 8. As the intensity of depolarizing current increased,
an inflection on the voltage trace appeared and grew in amplitude.
This inflection in the voltage trace is seen only with depolarizing
stimuli and not with hyperpolarizing stimuli. The time derivative
trace made this inflection even clearer, so that the rate of change
showed 2 peaks; the second peak indicated an active regenerative
component in the membrane potential change, which is characteristic
of active responses. The regenerative change was graded, perhaps
because of a large increase in K conductance.

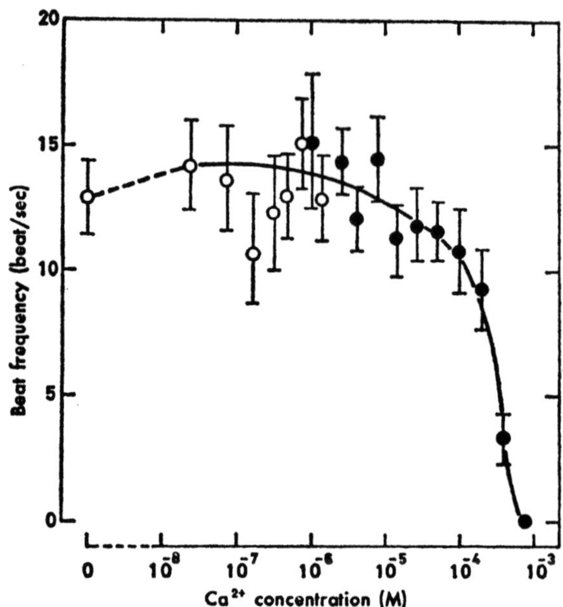

Figure 7. Beat frequency of cilia on reactivated models of _Parame-
cium_ as a function of Ca concentration. Meaning of solid and open
circles is the same as in Figure 6. From Naitoh and Kaneko (1972).

    Some of the properties of the membrane response are seen in
Figure 9. Hyperpolarization of the membrane reduced the amplitude
of the overshoot and the regenerative component. The regenerative
response was abolished by conditioning depolarizing pulses and
had an absolute refractory period of 30 msec. The response was
dependent on extracellular Ca (Figure 10). The overshoot increased
with a slope of 25 mV/10 fold increase in extracellular Ca. The
resting membrane was permeable to Ca but depolarization increased
the membranes' permeability to Ca so that the ideal 29 mv slope of
a Ca electrode was approached. These and other results suggested
that the membrane of _Paramecium_ underwent voltage-dependent increases in permeability to Ca which resulted in inward Ca current.
Indeed, a regenerative change in membrane potential was necessary
for the production of ciliary reversal in _Paramecium_ as reported
by Machemer and Eckert (1972).

    If the membrane regulated Ca influx, then setting the membrane
potential (Vm) near the Ca equilibrium potential should reduce Ca
influx and prevent ciliary reversal. The equation $I_{Ca} = g_{Ca}$ (Vm −
E_{Ca}) indicates the relations between Ca current (I_{Ca}), membrane
potential (Vm), and Ca equilibrium potential (E_{Ca}) according to the
ionic hypothesis (Hodgkin, 1964). As Vm approaches E_{Ca} and reduces
the driving force of Ca, then I_{Ca} should become very small and

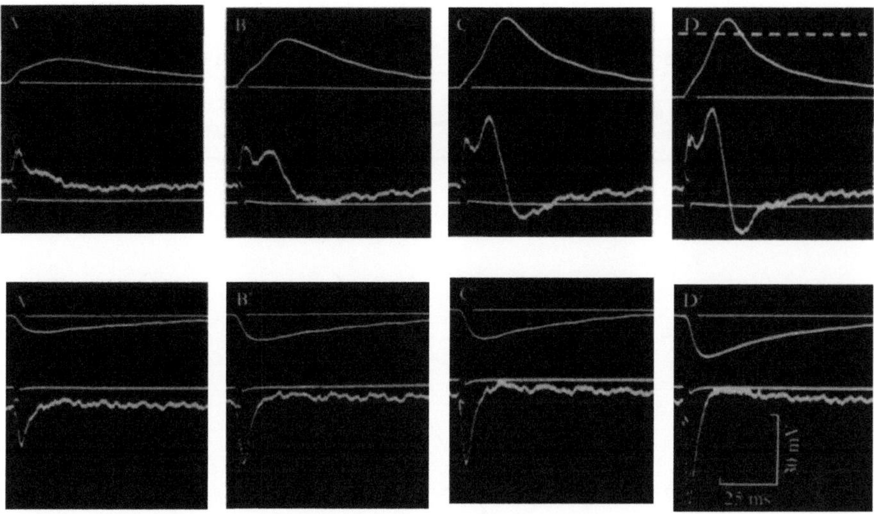

Figure 8. Membrane responses of <u>Paramecium</u> to increasing intensities of depolarizing (A-D) and hyperpolarizing (A'-D') current.
Upper trace, membrane potential; second trace, first time derivative
of Vm (dVm/dt); lowest trace, stimulating current intensity.
Sensitivity of dVm/dt is 2.5 V/sec for the height of the vertical
calibration line in D'. Dashed line in Figure D indicates zero
potential. From Naitoh, Eckert and Friedman (1972).

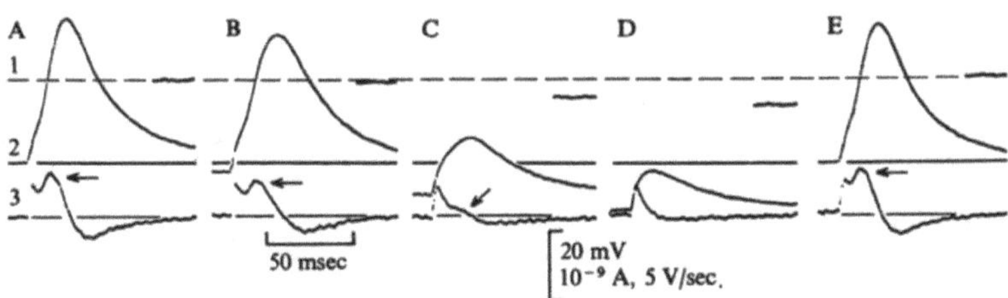

Figure 9. Reduction of response by hyperpolarization. Record A
shows the membrane response to a 2 msec depolarizing stimulus
applied at the resting potential. The resting level was hyperpolarized in progressive steps (B-D) after which the 2 msec stimulus
was applied. Record E shows the response after the end of the
hyperpolarizing steps. Trace 1 is the reference level for membrane
potential (Vm) and the deflection indicates the intensity of hyperpolarizing current; trace 2 is Vm· trace 3 is dVm/dt. Arrows in
trace 3 point to inflections indicative of regenerative component
in trace 2.

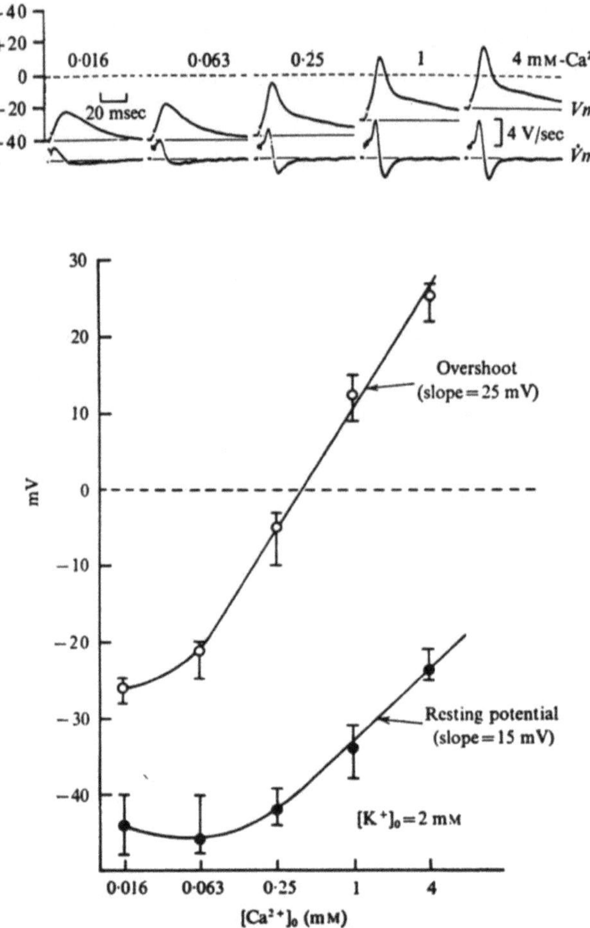

Figure 10. Membrane response to 2 msec current pulse is shown as
a function of extracellular calcium concentration. Potassium remained constant at 2m M as calcium was varied. Examples of responses
showing the membrane potential (Vm) and time derivative (V̇m) are
found at top of figure. Each point is the mean of 5 values with
deviations. From Naitoh, Eckert and Friedman (1972).

ciliary reversal should not result. In fact, when the membrane
potential of <u>Euplotes</u> is raised to a value of +70 mV, ciliary
reversal is suppressed for the duration of the current pulse as
illustrated in Figure 11. At the end of the current pulse a brief
"off" response or period of reversed beating occurred as the membrane
potential was repolarized. As a result of repolarization the driving force (Vm - E_{Ca}) increased and Ca entered the cell in sufficient
concentration to produce reversal. The observed "suppression"

potential of +70 mV was consistent with a calculated E_{Ca} of 116 mV, assuming a Ca concentration inside the cell of 10^{-7}M and outside of 10^{-3}M. Similar observations were made on the ciliates <u>Opalina</u> (Naitoh, 1958) and <u>Paramecium</u> (Machemer and Eckert, 1972).

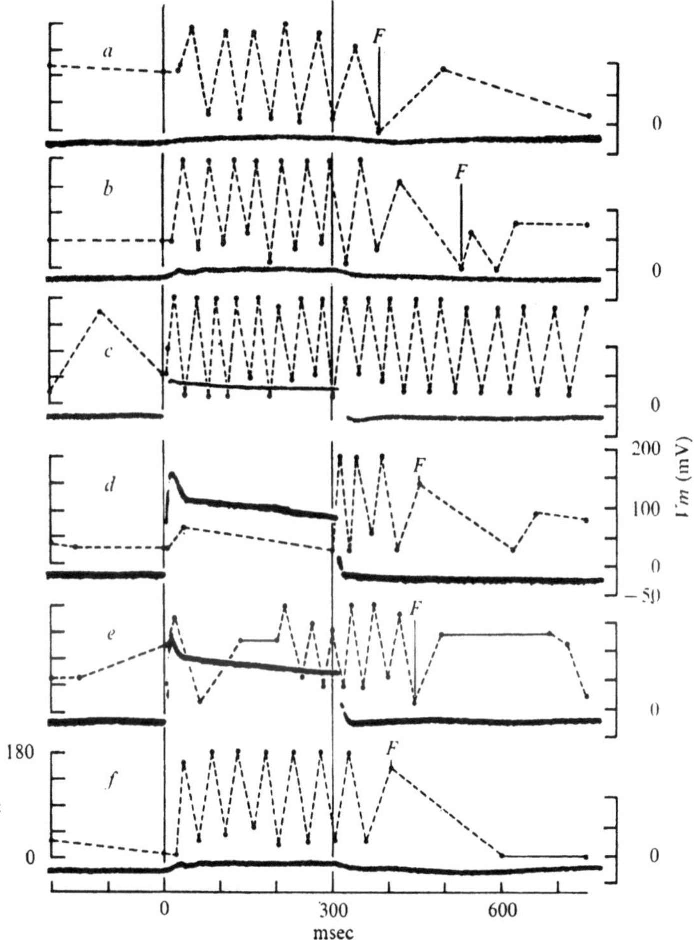

Figure 11. "Suppression of the reversed beating response in a single cell with large positive shifts in membrane potential. A series of increasing internally positive polarizations resulted in ciliary activity, which is plotted as extremes of inclination angle (α). Beating was in reverse except when the direction was forward as indicated by the letter F. The number of beats in reverse increased with larger positive shifts. In sequence d, however, no reversed beating occurred during the shift in membrane potential to +87 mV (value at the end of the pulse). Reversed beating was initiated with repolarization after the end of the stimulus pulse. Smaller potential shifts (sequences e and f) were accompanied by ciliary reversal. The resting potential remained at about -25 mV as indicated by the voltage scale at the right side of sequence d." From Epstein and Eckert (1973).

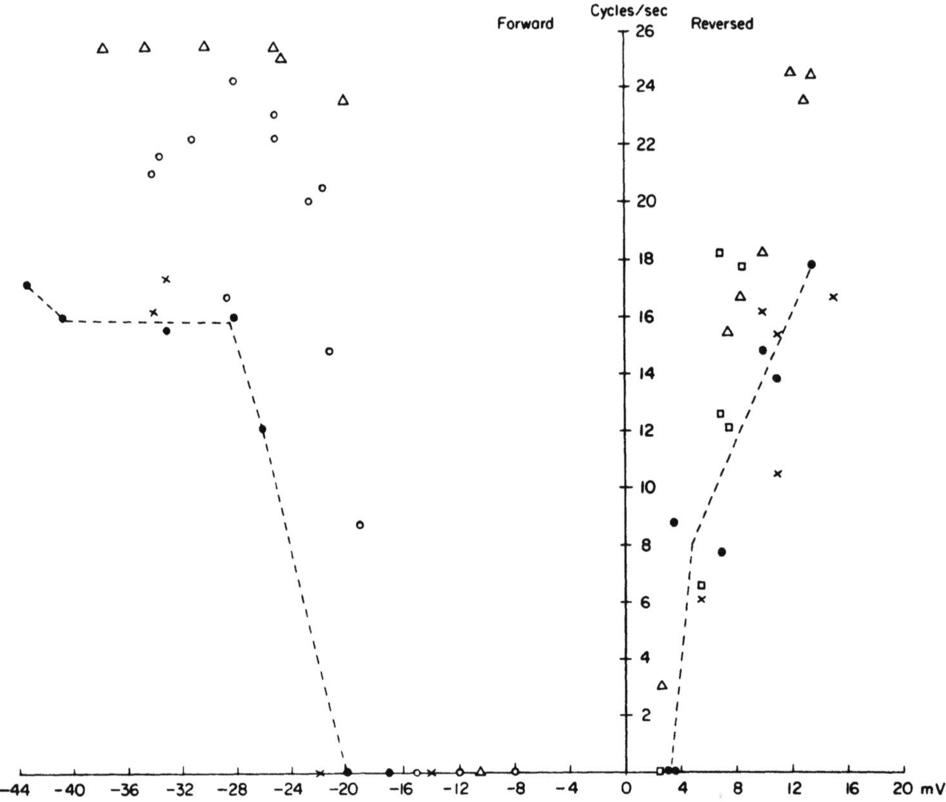

Figure 12. "The number of evoked cycles of beating of anal cirri
plotted against the steady-state shift in membrane potential produced by injected currents of 300 msec duration. Zero on the
abscissa is the resting potential. Depolarizing changes, indicated
on the right side of the abscissa, resulted in cycles of reversed
beating. Hyperpolarizing changes, shown on the left side of the
abscissa, evoked beating in a forward direction. Different symbols
represent a single cirrus on different cells. The dashed line
connects the values recorded for a particular specimen." From
Epstein and Eckert (1973).

Now let us consider the control of beating frequency. As
indicated above, large increases in frequency resulted from either
depolarizing or hyperpolarizing changes in membrane potential
(Figure 12). It has been found that only ATP and Mg are required
for beating of extracted models of flagella (Hoffman-Berling, 1955)
and cilia (Gibbons, 1965). Beating frequency of these models
varied with the concentration of ATP (Gibbons and Gibbons, 1972;
Eckert and Murakami, 1972). How then do changes in membrane
potential lead to changes in beating frequency? One possibility
for bioelectric control of frequency is Eckert's proposal (1972)
that Ca influx also stimulated metabolism causing an increase in
ATP concentrations. He suggested that the Ca concentration for
ciliary reversal and metabolic stimulation were different, with a
higher Ca concentration necessary for reversal. In the case of
hyperpolarization, the electrochemical gradient ($Vm - E_{Ca}$) for Ca
would increase, perhaps allowing a small influx of Ca which might
be sufficient to increase frequency but not sufficient to produce
ciliary reversal. The evidence for this hypothesis was that beating in _Paramecium_ (Eckert, 1972) and oviduct (Eckert and Murakami,
1972) stopped in a solution made low in Ca with EGTA. However,
this proposal requires further experimentation.

In summary, ciliary activity in many aneural organisms is under
membrane control. Appropriate stimuli result in an increase in the
membrane conductance, which in turn leads to an influx of Ca. The
Ca influx alters the orientation of beating and perhaps the frequency of beating.

ACKNOWLEDGEMENTS

The author thanks Dr. J.D. Sheridan for his comments on the
manuscript. This work was supported in part by U.S.P.H.S. Training
Grant 5T01 GM00448 to the Dept. Physiology., U.C.L.A. and U.S.P.H.S.
grant NS08364 and NSF grant GB- 30499 to Dr. Eckert, Dept. Zoology,
U.C.L.A., Los Angeles, California.

REFERENCES

Eckert, R. 1972. Bioelectric control of ciliary activity. _Science_,
    _172_, 473-481.
Eckert, R. and Murakami, A. 1972. in _Contractility of Muscle and
    Related Processes_, ed. R. Podolsky, Prentice-Hall, Englewood
    Cliffs, N.J.
Eckert, R. and Naitoh, Y. 1970. Passive electrical properties of
    Paramecium and problems of ciliary coordination. _J. Gen. Physiol._,
    _55_, 467-483.
Epstein, M. and Eckert, R. 1973. Membrane control of ciliary activ-
    ity in the protozoan _Euplotes_. _J. Exp. Biol._, _58_, 437-462.
Galt, C.P. and Mackie, G.O. 1971. Electrical correlates of ciliary
    reversal in _Oikopleura_. _J. Exp. Biol._, _55_, 205-212.
Gibbons, G.H. and Gibbons, I.R. 1972. Flagellar movement and
    adenosine triphosphatase activity in sea urchin sperm extracted
    with Triton X-100. _J. Cell Biol._, _54_, 75-97.
Gibbons, I.R. 1965. Reactivation of glycerinated cilia from
    Tetrahymena pyriformis. _J. Cell Biol._, _25_, 400.
Gliddon, R. 1966. Ciliary organelles and associated fibre systems
    in _Euplotes eurystomus_ (ciliata, hypotrichida) I. Fine Structure.
    J. Cell Sci., _1_, 439-448.
Hodgkin, A.L. 1964. _The Conduction of the Nerve Impulse_. Charles
    Thomas, Springfield.
Hoffmann-Berling, H. 1955. Geisselmodelle und Adenosintriphat.
    Biochem. Biophys. Acta, _16_, 146-154.
Kinosita, H. 1954. Electric potentials and ciliary response in
    Opalina. _J. Fac. Sci. Toyko Univ. Sect. IV._, _7_, 1-14.
Kinosita, H. and Murakami, A. 1967. Control of ciliary motion.
    Physiol. Rev., _47_, 53-82.
Machemer, H. and Eckert, R. 1972. Membrane calcium response and
    ciliary activity in _Paramecium_. _J. Cell Biol._, _55_, 161a.
Naitoh, Y. 1958. Direct current stimulation of _Opalina_ with intra-
    cellular microelectrode. _Annotnes. Zool. Japon._, _31_, 59-73.
Naitoh, Y. 1966. Reversal response elicited in nonbeating cilia
    of _Paramecium_ by membrane depolarization. _Science_, _154_, 660-662.
Naitoh, Y. 1968. Ionic control of the reversal response of cilia
    in _Paramecium caudatum_: A calcium hypothesis. _J. Gen. Physiol._,
    _51_, 85-103.
Naitoh, Y. and Eckert, R. 1969. Ciliary orientation: controlled by
    cell membrane or by intracellular fibrils? _Science_, _166_, 1633-1635.
Naitoh, Y., Eckert, R. and Friedman, K. 1972. A regenerative cal-
    cium response in _Paramecium_. _J. Exp. Biol._, _56_, 667-687.
Naitoh, Y. and Kaneko, H. 1972. ATP - Mg - reactivated Triton-
    extracted models of _Paramecium_: Modification of ciliary movement
    by calcium ions. _Science_, _172_, 523-524.
Roth, L.E. 1956. Aspects of ciliary fine structure in _Euplotes
    patella_. _J. Biophys. Biochem. Cytol._, _2_ (Suppl.), 235-240.

Roth, L.E. 1957. An electron microscope study of the cytology of the protozoan _Euplotes_ _patella_. _J_. _Biochem_. _Biophys_. _Cytol_., _3_, 985-1000.

Yamaguchi, T. 1960. Studies on the modes of ionic behavior across the ectoplasmic membrane of _Paramecium_ I. Electric potential differences measured by the intracellular microelectrode. _J_. _Fac_. _Sci_. _Tokyo_ _Univ_. _Sect_. _IV_, _8_, 573-591.

MEMBRANE POTENTIAL AND BEHAVIOR

PROPOSAL OF A MODEL SYSTEM

Victor Kai-Hwa Chen*
Department of Biophysics
Michigan State University
East Lansing, Michigan

I.   INTRODUCTION

Behavior has become one of the major interests of biology
today. As it relates to the mechanisms of the human brain, behavior
has been considered the ultimate challenge of man's intellect. The
search for the mechanisms of behavior has in the past been focused
on the vertebrate nervous system. As the result of the degree of
understanding that has been achieved through the study of invertebrate
organisms, there is a renewed interest in simple nervous systems as
the starting point of such a search. We are truly still at the
beginning. We have not achieved an understanding of the fundamental
mechanisms of behavior for any animal.

Neurobiology has become the new meeting ground for disciplines
seeking to understand the basis of animal behavior in terms of
cellular and subcellular events. These disciplines have come together
because of the profound changes which have occurred in our concepts
of neuronal information processing capabilities. I propose that this
new field should focus its attention upon a model system for behavior.
The model system should be free from constraining terminologies of
the individual disciplines, but extend the techniques of the
individual disciplines towards an understanding of the logic behind
behavior. The model system should have sufficiently intriguing
behavior and yet be approachable with forseeable techniques. I will
discuss how the ciliated protozoa could meet the requirements of
such a model system. The behavior of a few specific ciliates will be
considered in terms of membrane potentials and conductivities.
Neuronal mechanisms have been studied primarily with these subcellular
parameters in mind.

*Present address: Department of Biology, Case Western Reserve
University, Cleveland, Ohio.

II. CHANGING CONCEPTS OF NEURONAL INFORMATION PROCESSING

A. Changing Concepts of the Neuron

The textbook description of the structural and functional
properties of the neuron is no longer adequate. The description of
the neuron as a cell with a branched dendritic tree on one side
which conducts only graded potentials and an axon on the other side
which conducts only all-or-none potentials, fits only some of the
cells in the nervous system. It is no longer meaningful to label a
cellular processes an axon merely because it conducts action
potentials since even dendrites of some neurons conduct action
potentials (Purpura, 1967). Even the concept that all neurons
generate action potentials no longer holds. In the vertebrate retina,
the bipolar, the horizontal and the amacrine cells appear to process
information only in terms of graded potentials (Werblin and
Dowling, 1969). Some neurons in the vertebrate cortex are completely
silent electrically (Schmitt, 1967). Morphologically axons do not
synapse only onto dendrites. Axons are found to synapse onto cell
bodies and even onto other axons. Dendrites are not the only
processed which branch extensively, axons have been found to do so
also. Nerve cells are not like switches but are like organisms
different in their morphologies and highly changeable in their actions.

Behavior is considered as the expression by effectors of the
information processed by the neuronal networks of the organism.
We are used to considering the neuron as merely a network element
in which excitatory and inhibitory synaptic potentials interact
algebraicly, as proposed by Sherrington. Until recently we have
not even considered the neuron as capable of spontaneous activities.

Developmentally, nerve cells are closely related to epithelial
cells. In Cniderians, the simplest level of organisms with a
nervous system, epithelial cells have been implicated in behavioral
co-ordination (reviewed by Mackie, 1970; Josephson, 1973). Mackie
(1965) obtained good evidence for the conduction of behaviorally
meaningful signals by epithelial cells. He recorded propagated
all-or-none impulses from the nerve free exumbrellar epithelium of
hydrozoan medusae. The implication of this conduction concept to
neurobiology has been discussed by Josephson (1973). We know that
in these simple organisms there are also epithelial muscle cells
which co-ordinate behavior. Until the discovery of epithelial
conduction in larvae of Xenopus laevis (Roberts and Sterling, 1971)
by intracellular recording the phenomenon was considered just another
unusual finding in simple organisms. The skin cells of late embryo
and young larvae Xenopus are shown to conduct behaviorally meaningful signals before the nervous system develops in the periphery.
Since such an aneural conducting system can exist in the early
developmental stages of a vertebrate, the capabilities might still
be present in the adult. If such an aneuronal conduction were

expressed, even under special conditions in the adult vertebrate, we would be forced to re-examine our basic concepts of behavioral co-ordination mechanisms.

B.   Need For a Model System

The development of molecular biology of E. coli and its phage as a model system has resulted in the development of techniques to measure many non-electrical properties of cells. In the study of nerve and muscle cells we want to know more about neurons than just their membrane characteristics. After all,almost any cellular process in a neuron may be reflected in the neuron's actions in the behavioral co-ordination of the organism. The choice of E. coli and its phage as a model system drew the study of replication together. The choice of the model system focused the research efforts towards the fundamental logic behind replication. We need a model of a generalized information processing cell.

In the past neurophysiologists have studied behavioral processes at the level of neuronal circuits, behavioral scientists at the level of whole organisms. In the case of a ciliate model system the two approaches will be brought together. Research will then be directed at single cells which are also whole organisms. The behavioral concepts will thus be considered in terms of cellular mechanisms. Stentor as a model system will bring the neurophysiologist and behavioral scientists together to consider the basic logic of behavior.

Since we are aware that neurons are far more complex in their activities than switches, we need an understanding of the capabilities of the single cell. We can consider the neuron as merely a single cell covered with receptor and effector regions, the synapses being the receptors and the synapses and the axon being the effectors. As the result of the theoretical work of McCulloch and Pitts, it has been thought that the behavior of neuronal networks could be described by a network composed of two valued switches. Ultimately, the animal could be represented by an enormous network of these switches. This notion of the neuron as a switch was supported by the earlier intracellular studies, e.g., depolarization and the hyperpolarization of the neuronal membrane which resulted in initiation or blocking of action potentials to be conducted by the axon to the neighboring neurons. Attention has been directed towards the transmission of the information between neurons. A major attempt has been made to map the neuronal interactions in the nervous system of many animals. Because of the complex information processing abilities of some neurons it has become necessary to carry the mapping to the electron microscope level (Waxman and Pappas, 1972). We are beginning to think of information processing in terms of circuits with synapses rather than cells as elements. Mapping the interneuronal connections may yet be insufficient.

It has been suggested in some adult animals that not all of the
morphologically identified synapses are used in a natural behavior
although the synapses can be stimulated electrically. Perhaps we
should begin to study what information processing could occur
between synapses on the neuronal surface.

III. A CILIATE MODEL OF BEHAVIOR

Electrophysiology of Ciliates

The concepts and methods which have been used to study ciliate
behavior are the same as those used to study single nerve cells.
The ciliated protozoan were among the first cells to be studied
with intracellular pipette electrodes. Gelfan (1927) measured the
cytoplasmic resistivity of several protozoa, including <u>Stentor</u> and
<u>Paramecium</u>, with pairs of intracellular electrodes. He used quartz
capillaries drawn to 1-2 micron diameter tips for his electrodes.
These intracellular electrodes were filled with 3M. KCl agar. Gelfan's
resistivity values were used recently for the calculation of
<u>Paramecium's</u> membrane length constant (Eckert and Naitoh, 1972).
It is of interest that intracellular recording of ciliates have
been made so long ago. We usually consider the study of cells other
than nerve cells and muscle cells by intracellular electrodes as a
very recent technique. It is only recently that this technique has
been applied to eggs, to glands and to epithelium. But among the
first cell membrane potential measurements were those performed by
Kamada (1931) who studied the membrane potentials of <u>Paramecium</u>.
Since that time the membrane potential of several other ciliates
have been studied (reviewed by Eckert and Naitoh, 1972). The study
of the electrical behavior of cells through the use of intracellular
electrodes has been one of the major techniques in neurobiology.
The mechanisms for locomotor behavior of ciliates have long been
considered in terms of membrane potentials and ionic conductances,
which are two parameters closely associated with the study of neural
function.

A. Reflex-like Behavior in Free Swimming Ciliates

Ciliated protozoa, though single cells, should be thought of
as complete behaving organisms. The contemporary protozoa should
not be considered primitive except in a phylogenetic sense. They
have evolved through the common environmental challenges along with
the metazoans. Rotifers, metazoa composed of approximately five
thousand cells, can often be found in the same drop of pond water
with the common ciliated protozoa, <u>Stentor</u>. The two animals are of
similar size. They both use cilia as their only means of locomotion.
<u>Stentor</u>, which is but a single cell, can often be observed to feed
upon a rotifer. Rotifers have nervous systems containing approximately two hundred nerve cells. The behavioral capabilities of the
sensory and co-ordination mechanisms of rotifers and <u>Stentor</u> appear

to be comparable. At least the mechanisms of the single-celled
Stentor are not clearly inferior to those of the rotifer which have a
nervous system. Jennings (1899) observed that the response of
the free swimming rotifer to stimulation agrees even in detail with
that of the ciliate. A single cell can develop information
co-ordinating and processing mechanisms to a degree of sophistication
comparable to that in a system of two hundred nerve cells.

From an evolutionary view, single-celled organisms are limited
in size by the physical law of diffusion rather than by their
co-ordinating abilities. The abilities of the single cell are
limited by its surface area and volume. In the metazoa, the cells
have specialized, yet they may be capable of expressing motility,
contractility and irritability under special conditions. In the
nervous system, in particular, nerve cells may express these general
cellular abilities under special conditions. Study of single-celled
ciliates will enable us to better decide whether information
processing in the metazoan nervous system is the result of the
pattern of neuronal connections or the result of individual
decision-making cells. The ciliate could be considered a model of
the neuron with most of its general cellular abilities expressed.

Of the ciliates, Paramecium has been most extensively studied
by electrophysiological techniques. The locomotor behavior of
Paramecium has been shown to be controlled by the membrane.
Researchers at the University of Tokyo and at UCLA have shown that
the membrane regulates the intracellular ionic concentrations which
control the activity of the effectors. They have shown that ciliates
are not so different electrophysiologically from the individual
receptor, neuron or muscle cells in the metazoan behavioral system.
Paramecium represents one of those ciliates which swim ceaselessly.
It does not have the variability in behavior of Stentor which can be
free swimming or attached. Nor does Paramecium exhibit the
variability in behavior of the attached ciliates such as Vorticella.

Free swimming ciliates typically spiral through the water like
a projectile. This is true even for Vorticella when it has broken
away from its attachment. As in the case of the projectile, the
rotation of the asymmetrical ciliate body results in an over-all
straight course (Jennings, 1906). Free swimming rotifers spiral in
their swimming in a similar manner (Jennings, 1899). In a given
ciliate species there is a predominate direction of rotation
(Tartar, 1961). The rotation of the body is due to the oblique
beating direction of the body cilia (Jennings and Jamieson, 1902).

When free swimming Stentor or Paramecium encounter a stimulus
it swims backward, turns toward the aboral side and then swims
forward again in a new direction. Stentor contracts as it swims
backwards. The response sequence is termed "avoiding reaction"
(Jennings, 1906). Most electrophysiological studies with

Paramecium have been on the avoidance response. Specifically, it has
been shown that it is the membrane which controls the reversal of
the ciliary beat direction (reviewed by Eckert and Naitoh, 1972).
In the "avoiding reaction" of S. polymorphus it has been shown that
body contraction occurs first and is then followed by cessation of
the membranelle beating and forward pointing of the membranelles.
The transient membrane potential change for the two steps are of
opposite polarity and occur sequentially (Chen, 1972). Although
Paramecium can contract (Nagai, 1956), the membrane potential
correlates have not been studied. It has been reported that
spontaneous membrane potential changes occur corresponding to the
transient stopping and forward pointing of the membranelles in
Stentor (Mergenhagen, 1971; Chen, 1972). The body contraction of
unattached Stentor can be stimulated mechanically or by depolarizing
the membrane (Wood, 1970; Chen, 1972). The form and beat of Stentor's
membranelles has been studied in great detail (Sleigh, 1956). The
free swimming Stentor and Paramecium share the stereotyped nature
of the reflex behavior of vertebrates. The reaction is present even
when the anterior half of Stentor is cut off (Jennings and Jamieson,
1902).

     Stentor is often seen to move over the bottom of the culture
dish with its oral disk applied to the surface of the dish, apparently
feeding (Tartar, 1961). When Stentor ceases revolving around its
long axis it bends and attaches its holdfast to the dish. After
attachment of the holdfast the oral disk points away from the surface
of the dish. The detailed studies of Andrews (1945) indicated
that the holdfast organ is a rapidly changing structure. The organ
is capable of forming cilia, pseudopodia and other structures for
attachment. The holdfast is drawn into the body somewhat when
Stentor is free swimming. It would indeed be of interest to look
for membrane potential changes or membrane currents which co-ordinate
the activities of this organ. Since the behavior of attached
Stentor is more variable than the free swimming animal, the holdfast
may be the behavioral switch which releases the variable behavior
when attached to the substrate.

B.   Intriguing Behavior of Attached Stentor

     Jennings (1902) described a series of responses of attached
Stentor roeseli to stimuli until the noxious stimulus was avoided.
He found that a light touch resulted in no response or a bending
towards the stimulus. If a cloud of carmine particles was dropped
on Stentor the animal twisted and then bent towards its aboral side.
This bending towards a structurally defined side is reminiscent of
the "avoiding reaction" but since the animal is bending it was not
necessarily removed from the stimulus. Another response to carmine
particles or similar stimuli was the transient stopping of the
membranelles with the transient reversal of the body cilia. This

transient reversal of the beating may be repeated several times.
If the stimulus persisted or if the stimulus was very strong the
animal contracted immediately. It extended slowly and re-contracted
if the stimulus persisted. This contraction and slow extension
could last for some time. Finally the holdfast detached and
Stentor swam away. Tartar (1961) found that Stentors from recently
fed cultures were more likely to remain attached and give the graded
response series than unfed animals. Reynierse and Walsh (1967) were
unable to demonstrate a graded response series in S. coeruleus.
The animals quickly detached and became free-swimming. Jennings (1902)
reported that attached Stentors sometimes accommodated to the
continued application of mechanical stimuli. They no longer
contracted to stimuli but responded in other ways. It was shown
that this failure to respond was not the result of sensory fatigue.
Jennings' description of the response is often reported in textbooks.
Perhaps because it has been repeated so often, there is little
analysis of it in the literature. The response sequence for attached
S. roeseli would be the ideal next step, since the stereotyped responses
of free swimming ciliates have been studied extensively. This
behavior could have significant value to neurobiology since it has
sufficient complexity to be of interest. Not only are there species
of Stentor which follow the response sequence to different degrees,
there are species and strains which do or do not respond to various
stimuli. For example S. coeruleus is easily detached, S. roeseli
is difficult to detach, and there are races of S. polymorphus
which are sensitive to daylight and others which are insensitive.
Food selection is another behavior which could have meaning to a
search for the fundamental logic of behavior.

We have so far only discussed the simple movements of Stentor
in response to stimuli. The proposed model system has even more
interesting behavior than these simple movements. Food selection
in the model system is such a behavior. Selection of food particles
by Stentor has been clearly demonstrated by Schaeffer (1910), who
found that localized reversal of cilia beating selectively rejected
some particles while others were ingested. The amount of the particular food ingested was shown to be dependent upon the presence of
other substances. Schaeffer found that Stentor was more dis-
criminating when it was satiated than when it was starved.
Hetherington (1932) reported that Stentor had a preference for
ciliates and was capable of selection even within this group. Even
when starved Stentor rejected autotrophs such as Gonium and Euglena.
In a recent report (Rapport et al., 1972) the food preference of
Stentor was carefully re-examined. Experiments were designed so
that food preference could be measured statistically without
confounding differences in "catchability" with choice. Their
findings are consistent with Hetherington's observation of the
preference of ciliates over algal prey. I propose that membrane
potential changes in Stentor could be followed during the food
selection sequence. The receptor potential indicating the

detection of the prey could be followed to the initiation of the
transient potential changes rejecting the prey by ciliary reversal.
This moderately complex behavior of food selection represents a
behaviorally relevant decision process accessible to study at the
cellular level with conventional neurobiological techniques.
Stentor is particularly suitable since the techniques such as
alteration of ionic conditions, exposures to drugs and chemical
dissection to alter behavior have already been developed.

III. SPECIAL METHODS FOR DISSECTING BEHAVIOR

A. Chemical Dissection of Behavioral Structures

Ablation of specific receptors and effectors has been an
important tool in the study of behavioral co-ordination in many
animals. Removal of specific structures allows the roles that the
structure plays in the servo loops to be ascertained. The alteration of the walking and the running co-ordination patterns in
insects have been extensively studied using this technique (reviewed
by Wilson, 1966). In cockroach, antenna cleaning has been studied
in this manner. The animal holds the antenna being cleaned with
the opposite foreleg. When both forelegs are removed the animal
learns to hold the antenna with a middle leg (Luco and Aranda, 1964).
This surgical technique is difficult except where the animal to be
studied has well defined structures. In the case of small organisms
such as ciliates surgical ablation lacks exactness at the cellular
level. Cutting and grafting techniques have been used to study
cellular regulation and differentiation in Stentor because of its
remarkable ability to survive and reorganize after such operations
(Tartar, 1960). The surgical technique though useful for large
animals with prominent structures, would lack the specificity desired
for behavioral studies with animals as small as Stentor.

Stentor has the unusual morphological characteristic that many
of its structures can be removed selectively by chemical treatment
(Tartar, 1961). Application of specific chemical solutions results
in the shedding of specific structures. Some solutions will result
in the shedding of several structures in a particular sequence.
The principle structures of interest to a behavioral study such as
the membranellar band, pellicle and pigment can all be chemically
removed. The specificity of this technique is dramatically
illustrated in the removal of the membranellar band by treatment
with sea water and Holtfreter s solution. In sea water the membranellar band is lifted first at the distal end. In Holtfreter's
solution the membranellar band lining the gullet is lifted off
first (Tartar, 1961). The opposite directions of the removal of
the membranellar band strongly suggest the presence of chemical
asymmetry in the structure. Behaviorally there is an asymmetry in
the two directions of the membranellar band, e.g., the metachronal

wave only passes along the band in one direction. Thus considerable
specificity can be achieved; the membranelle plates can even be
removed without shedding of the underlying structures of the
membranellar band. Tartar (1968) has used this method to
demonstrate that cilia can beat before they have regenerated to
their full length. Chemical dissection is an ideal technique for
behavioral studies in that the structures are cleaved with
specificity at the molecular level. This method allows many
animals to be treated simultaneously thus allowing massed behavioral
testing. Such a method would also provide sufficient material
for analysis of the chemical nature of the structures involved.
This morphological characteristic of _Stentor_ will allow the co-
ordination mechanisms to be studied to a new degree of refinement
with the structural ablation technique.

When _S. polymorphus_ is current clamped, to transiently stop
membranellar beating, a "hump" is often seen in the transmembrane
potential waveform. It was suspected that the "hump" was due to
the membranelle beating. Treating the animals in sucrose, it was
possible to selectively shed the membranelles or the complete
membranellar band (Tartar, 1968) It was found that animals treated
in this manner still could develop the "hump" in their membrane
potential waveforms, thus suggesting that the characteristic of the
waveform was not produced by membranellar band activity (Chen, 1972)
Using the chemical dissection technique one could perhaps assign
particular membrane currents or membrane conductivity changes to the
behavior of specific structures. It is not difficult to see how
this technique could be used to delineate the behavioral pattern of
food selection in terms of membrane changes.

B. Ionic Dissection of Co-ordination Mechanisms

One of the major tools in neurobiology used to study cellular
mechanisms has been to alter ionic conditions. This powerful approach
has been used to study action potentials generated in the single
neuron. Since _Stentor_ is both a single cell as well as a complete
behaving organism the use of altered ionic conditions to identify
the ionic currents responsible for the membrane potentials would
also identify the ions which activate specific effectors.

The metachronal beating of the membranelles which surround the
oral disc of _Stentor_ has long fascinated observers. Sleigh (1956,
1957) has studied this metachrony in great detail through alterations in the bathing medium. Sleigh (1957) has shown that
hydrodynamic coupling of the beating membranelles was insufficient
to explain the observed metachrony. In low concentration of $AlCl_3$
or $MgCl_2$ he found that the frequency of the metachronal wave
increased without changing the wave velocity. (Hydrodynamic
coupling of beating cilia has been proposed to be the mechanism

for the metachronal beating of cilia on <u>Paramecium</u>, however.)
The co-ordination mechanisms of the membranellar band was dissected
by Tartar (1957). Using much higher concentrations of $MgCl_2$ than
Sleigh, Tartar found that the membranelles started and stopped at
the rate of one change per second. In very weak $NiSO_4$, Tartar
observed that although the membranelles were not beating they moved
"like the batting of eyelashes". In $CaCl_2$ the membranelles beat
vigorously without reversal. It appears that the mechanisms for
co-ordination are distinct from those for beating and reversal.

As in the case of the mechanisms controlling the membranelles,
the control of the body cilia appear also to be dissectable with
altered ionic conditions. The primary effect on the body cilia by
various ions is the alteration of ciliary reversal. In distilled
water, <u>Stentor</u> swims backward continuously. In KCl, ciliary
reversal, metachronal rhythm and ciliary beating are lost successively,
indicating that the three types of ciliary actions are separate
processes. It is of interest that Ca^{++} and Mg^{++} do not induce
ciliary reversal of the body cilia in <u>Stentor</u> (Tartar, 1961).

The effect of chemicals which cause <u>Stentor</u> to contract has
not been as extensively studied as those producing ciliary reversal.
Tartar (1957) found that animals treated in NaI or Kl could be cut
in two without contracting. The blocking of contraction by NaI and
Kl was completely reversible. I have found that <u>S. polymorphus</u>
treated with Kl would still contract to intracellular current
injection though the threshold for contraction to mechanical
stimulation was greatly increased. The effect of Kl seems to be
upon the receptor mechanisms rather than the contraction mechanisms
(Chen, 1972). Neresheimer (1907) used drugs which affect a nerve-
muscle preparation to study contraction of <u>Stentor</u>. He found that
the insensitivity produced by morphine hydrochloride was counter-
acted by picrotoxin and atropine. Strychnine and curare produced
strong contractions. The action of curare could not be blocked by
physostigmine. Contraction in <u>Stentor</u> has been thought to be due to
bundles of fibers which resemble microtubules. These bundles are
known as myonemes. <u>Stentor</u> could be used as a model to study
excitation-contraction coupling.

Altering ionic conditions to study the behavioral mechanisms of
<u>Stentor</u> will lead to the development of useful ideas which could be
applied to our study of nerve and muscle cells. Since the drugs
tested by Neresheimer have similar effects in different organisms,
it might be suggested that at the chemical level, nature may not
have employed many different solutions to control behavior.

IV. IMPLICATIONS OF THE MODEL SYSTEM

Our knowledge of the abilities of nerve cells have come
primarily from detailed studies of model systems. A model system
serves to simplify the variables which we want to analyze. An
important assumption behind the use of a model system is that the
variables of interest interact in the model in the same manner as
they do in more complex systems. Most of what we know about action
potential conduction is based upon the giant axon of the squid.
It is an axon resulting from the fusion of processes of many nerve
cell bodies. This is of special interest since the cellular view of
the nervous system has long been predominant over the reticular view.
It is only recently that a certain degree of continuity between
neurons has been found in the vertebrate nervous system. It requires
a certain degree of faith that the properties of a fused axon should
have the same properties as the axon from a single cell body. The
giant cell bodies of Aplysia have allowed us to study neuronal
circuits responsible for some basic behaviors. These giant cell
bodies have also shown us that slow and long term membrane changes
are as important in determining behavior as rapid action potentials.
Another important idea that we have learned from these cells is that
membrane properties can vary greatly over small distances on
seemingly uniform areas of membrane (Neher and Lux, 1969). These
patches have quite different ionic currents. This property of
membranes is perhaps amplified on the surface of the single-celled
ciliate. On the surface of an animal such as _Stentor_ many
information processing events must be occurring simultaneously.
Stentor seems to be the model system we can use to study the limits
achievable at the single cell level. The discovery of neurons
capable of extensive information processing in the vertebrate
cortex, suggest that a model system such as _Stentor_ is needed.

Stentor as a model system will enable us to study the logic of
behavior which is present in the unicellular organism as well as
animals with a well developed nervous system. The free swimming
rotifer with a nervous system of approximately two hundred nerve
cells has been observed to respond to stimulation in the same
manner as the free-swimming ciliate. Typically both the ciliate
and the rotifer spiral through the water to maintain a straight
course. Considered from this point of view the neuron does not
look to be different from the single-celled ciliate. The ciliate
also has its effectors and receptors limited to the cell surface.
The ciliate would be a model of a general information processing
cell. We are used to considering the nerve and muscle cells as
having lost many general cellular abilities such as movement and
cytoplasmic streaming. Cytoplasmic streaming in nerve cells is
now a phenomenon of major interest. In culture the nerve and
muscle cells can still express the general cellular property of
movement. Perhaps _in vivo_ the restraints are not as rigid as
previously thought. Recent studies in _Daphnia_ showed that some

variability is still possible in neuronal connections even when
comparisons were made between isogenic individuals (Macagno, et.al.,
1973). In the ciliate model system we should develop a clear concept
of the limit of information processing achievable by a single cell.

The correlation of behavior with membrane potentials has been
a very fruitful approach in the study of the individual cells in
metazoa. The relation of this same parameter with the behavior of
the single-celled ciliate would yield new insight on the role of
membrane potentials in cellular information processing. Membrane
potential variation can be considered as one of the universal
symbols in the basic logic of behavior. The variation of its value
alters the expression of behavior in single-celled ciliates as well
as other animals. Certainly there are behavioral phenomena at the
metazoan level not present at the protozoan level. Using nature's
experiments, we may be able to learn her universal symbols.
Understanding the system logic for the behavior of a simple
organism will give insight as to the logic she may have used at
other levels of behavioral complexity.

Aside from the grand quest to understand the fundamental logic
of behavior, the ciliate model system can also provide answers of
immediate interest to neurobiology. Axoplasmic streaming is a
cellular process of current interest. The ciliate _Tokophyrya_ is an
excellent model system showing the movement of cytoplasm by microtubules (Rudzinska, 1973). The contraction of _Stentor_ is often
thought to involve microtubules. The chemotaxis of bacteria as a
model for receptor behavior is currently of great interest (reviewed
by Doetsch, 1972). We should note that _Stentor_ exhibits chemotactic
behavior to a variety of ions in solution (Pietrowicz-Kosmynka, 1971).
The food selection behavior no doubt involves the use of chemoreceptors
for discrimination of the food particles. The _Stentor_ model system
would serve as the next step from the study of bacterial chemoreception.

I propose that the use of _Stentor_ as a model system for the
study of behavior will serve to unify the efforts of many disciplines.
The use of the _Stentor_ system will at the same time promote the
search for the fundamental logic of behavior. Since the model has
no nervous system it will allow the search for a universal logic of
behavior in terms of general subcellular mechanisms.

ACKNOWLEDGEMENTS

I thank R.K. Josephson and May Lee for critical comments on
this chapter.

REFERENCES

Andrews, E.A. 1945. _Stentor's_ anchoring organs. _J. Morph._, _77_,219-232.

Chen, V.K. 1972. The electrophysiology of _Stentor polymorphus_:
  An approach to the study of behavior. Ph.D. thesis, State Univ.
  of New York at Buffalo, Dept. of Biophysics.
Doetsch, R.N. 1972. A unified theory of bacterial motile behavior.
  J. Theor. Biol., _35_, 55-66.
Eckert, R. and Naitoh, Y. 1972. Bioelectric control of locomotion
  in the ciliates. _J. Protozool._, _19_, 237-243.
Gelfan, S. 1927. The electrical conductivity of protoplasm and a
  new method of its determination. _Univ. of Calif. Pub. in Zool._,
  _29_, 453-465.
Hetherington, A. 1932. The constant culture of _Stentor coeruleus_.
  Arch. Protistenk., _76_, 118-129.
Jennings, H.S. 1899. Reaction to stimuli in certain Rotifera.
  Carnegie Inst. of Wash. Pub., _16_, 75-87.
Jennings, H.S. 1902. Studies on reactions to stimuli in unicellar
  organisms. IX, on the behavior of fixed infusion (_Stentor_ and
  Vorticella) with special reference to the modifiability of
  protozoan reactions. _Amer. J. Physiol._, _8_, 23-60.
Jennings, H.S. 1906. _Behavior of the lower organisms._ Columbia
  Univ. Press, New York.
Jennings, H.S. and Jamieson, C. 1902. Studies on reactions to
  stimuli in unicellular organisms. X, the movements and reactions
  of pieces of ciliate infusions. _Biol. Bull._, _3_, 225-234.
Josephson, R.K. 1973. Cnidarian Neurobiology. To appear in
  Perspectives in Coelenterate Biology, Lenhoff, H. and Muscatine, L.
  (Eds.), Acad. Press, New York.
Kamada, T. 1931. Reversal of electric polarity effect in _Paramecium_
  according to the change of current strength. _J. Fac. Sci. Imp._
  Univ. Tokyo, sec IV, _2_, 299-307.
Luco, J.V. and Aranda, L.C. 1964. An electrical correlate to the
  process of learning. Experiments in _Blatta orientalis._ _Nature_
  _209_, 205-206.
Macagno, E.R., V. Lopresti, and C. Leventhal 1973. Structure and
  Development of Neuronal Connection in Isogenic Organisms:
  Variations and Similarities in the Optic System of _Daphnia magna._
  Proc. Nat. Acad. Sci. USA, _70_, 57-61.
Mackie, G.O. 1965. Conduction in the nerve-free epithelia of
  siphonophores. _Amer. Zool._, _5_, 439-453.
Mackie, G.O. 1970. Neuroid conduction and the evolution of con-
  duction tissues. _Quart. Rev. Biol._, _45_, 319-332.
Mergenhagen, D. 1971. Membrane potentials in _Stentor coeruleus._
  Protoplasma, _72_, 359-365.
Nagai, T. 1956. Elasticity and contraction of _Paramecium_
  ectoplasm. _Cytologia_, _21_, 65-80.
Neher, E. and Lux, H.D. 1969. Voltage clamp on _Helix pomatia_
  neuronal membrane: current measurement over a limited area of the
  soma surface. _Pflügers Arch._, _311_, 272-277.
Neresheimer, E. 1907. Nochmals über _Stentor coeruleus._ _Arch._
  Protistenk., _9_, 137-138.

Pietrowicz-Kosmynka, D. 1971. Chemotactic effects of cations and
   pH on Stentor coeruleus. Acta Protozoologica, 9, 235-244.
Purpura, D.P. 1967. Comparative physiology of dendrites. In The
   Neurosciences, Quarton, G.C., Melnechuk, T., and Schmitt, F.O.
   (Eds.), The Rockefeller Univ. Press, New York, p. 372-392.
Rapport, D.J., Berger, J., and Reid, D.B.W. 1972. Determination of
   food preference of Stentor coeruleus. Biol. Bull., 142, 103-109.
Reynierse, J.H. and Walsh, G.L. 1967. Behavior modification in
   the protozoan, Stentor, re-examined. Psychological record, 17,
   161-165.
Roberts, A. and Sterling, C.A. 1971. The properties and propaga-
   tion of a cardiac-like impulse in the skin of young tadpoles.
   Z. Vergl. Physiologie, 71, 295-310.
Rudzinska, M.A. 1973. Do suctoria really feed by suction.
   Bioscience, 23, 87-94.
Schaeffer, A.A. 1910. Selection of food in Stentor coeruleus.
   J. Exp. Zool., 8, 75-132.
Schmitt, F.O. 1967. Molecular neurobiology in the context of the
   neurosciences. In The Neurosciences, Quarton, G.C., Melnechuk, T.,
   and Schmitt, F.O. (Eds.), The Rockefeller Univ. Press, New York,
   p. 209-219.
Sleigh, M.A. 1956. Metachronism and frequency of beat in the
   peristomial cilia of Stentor. J. Exp. Biol., 33, 15-28.
Sleigh, M.A. 1957. Further observations on co-ordination and the
   determination of frequency in the peristomial cilia of Stentor.
   J. Exp. Biol., 34, 106-115.
Tartar, V. 1957. Reactions of Stentor coeruleus to certain
   substances added to the medium. Expt. Cell Res., 13, 317-332.
Tartar, V. 1960. Reconstitution of minced Stentor coeruleus.
   J. Expt. Zool., 144, 187-207.
Tartar, V. 1961. Biology of Stentor., Pergamon Press, New York.
Tartar, V. 1968. Regeneration in situ of membranellar cilia in
   Stentor coeruleus. Trans. Amer. Microsc. Soc., 87, 297-306.
Waxman, S.G. and Pappas, G.D. 1972. Changing concepts of the
   neuron. Microstructures, 3(2) pp. 13-16 and 25.
Wilson, D.M. 1966. Insect walking. Ann, Rev. Entomol. 11,103-122.
Werblin, F.S. and Dowling, J.E. 1969. Organization of the
   retina of the mudpuppy, Necturus maculosus. II,Intracellular
   recording. J. Neurophysiol., 32, 339-355.
Wood, D. 1970. Electrophysiological correlates of the response
   decrement produced by mechanical stimuli in the protozoan,
   Stentor. J. Neurobiol., 2, 1-11.

CONTRACTILITY OF MUSCLE CELLS AND

NON-MUSCULAR CONTRACTILE CELLS

Earl M. Ettienne
Department of Anatomy
Harvard Medical School
Boston, Massachusetts

To fully understand the roots of our current concern over the
application of biological techniques to behavior, one needs to
consider the pioneering work of Darwin, Pavlov and Mendel. The
works of each of these men provided radical insights into comparative aspects of behavior in different species; physiological
responses to behavioral conditioning and a methodology for understanding the hereditary basis for behavioral traits in living organisms.

It is primarily through the insights provided by these early
investigators that we began to understand that behavioral characteristics such as avoidance, predation, reproduction, learning
etcetera, are common throughout the living world.

While our understanding of the general physiological and
mechanical principles which jointly govern the organization of higher
organisms is considerably sophisticated today, it was only through
an elaboration of Mendelian genetics that we began to understand
the processes through which nature selects and faithfully reproduces
those characteristics in living organisms which are most suitable
for their survival in a particular environment.

The attempt to understand the general processes of selection
in response to changing environmental conditions and the maintenance
of the specific selective response may have given rise to the major
disciplines in both the behavioral and natural sciences.

Modern genetics attempts to explain the processes underlying
selection and reproduction of the selective response, by understanding the molecular events which accompany observable changes within

a living system. Modern genetics also uses biochemical tools to
understand how selective information is "stored", "transcribed"
and effectively "translated" in living systems. It is also
possible to observe with biochemical tools the effect of a
"translated" message on the dynamics of a living system. The
developments in genetics represent considerable advances in scien-
tific methodology in a field that was originally comparative or
descriptive in methods of study. Similar methodological advances
occur in cellular and developmental biology, neurobiology and
molecular biology. The behavioral sciences provide statistical
data on the observed response(s) of whole organisms or groups of
organisms to specific environmental conditions which may be controlled by the behaviorist.

While these are necessarily limited definitions, they point
to ways in which the logic inherent in each of the fields listed
above may be interrelated.

I. ELECTROCHEMICAL CONDUCTION AND EXCITATION

Neuroanatomists and physiologists have given us an understanding of the neurological organization of the somewhat complex
multicellular invertebrate systems (Kandel, 1970) to the more
complicated vertebrate systems. These observations provide
insight into how sensory stimuli are relayed through the nervous
system. The general understanding is that an excitatory stimulus
effects a reversal in an electrical potential maintained at the
expense of metabolic energy across the neuronal membrane. The
depolarization is then propagated electrogenically along the
membrane by the reversal of permeability of the membrane to Na^+
and K^+ ions. While other ions have been implicated in the process
of membrane-associated electrical conduction, it is generally
accepted that Na^+ and K^+ are the principal ions involved in the
observed phenomenon (Hodgkin, 1951).

Neurons communicate with each other and with other cells
through electrochemical, cell-cell, junctions. At these junctions
or synapses, an arriving electrical signal triggers the release
of a chemical transmitter. The synaptic neurotransmitters may be
of two types: excitatory, such as nor-adrenaline or acetylcholine;
or inhibitory, such as dopamine and 5-hydroxytryptamine.

The organization of neuronal networks and the transmitter
properties of the synapse are thought to reflect the integration
of excitatory and inhibitory stimuli and thus affect a complex
pattern of behavior in an organism.

In thinking of living organisms as neurological systems which
are responsive to stimulation, excitability must involve a sensitive

receptor and a response mechanism. The response may take the form
of a specific behavior or reaction different in nature from the
triggering event. The response to a stimulus applied to an organism might be a specific set of muscle contractions which may
ultimately be interpreted as a behavioral response specific to the
stimulus for that particular organism. Thus, excitability must
involve a number of separable, though integrated events. The
stimulus must first interact with a "receptor". The receptor must
then transfer the signal, generally through the electrochemical
process associated with nerve conduction, to an "effector". The
effector produces a response which may be characteristic for the
stimulus. Receptors and effectors in multicellular organisms
often exist as or in separate cells. The situation becomes
radically different in single-celled systems which are capable of
a wide variety of effector reactions in response to mechanical,
electrical, chemical, photic and thermal stimulation. The organ-
ismic responses may be as varied as the stimuli.

In observing behavioral responses to stimuli, it becomes
apparent that biological movement (motility) is a process fundamental to most systems. The observation is valid whether we are
concerned with populations, multicellular systems, or single cells.
The motile behavior may be connected to stimulation to feed,
reproduce, avoid, circulation and respiration and may occur at the
level of organelles. Motility in general results from the action
of specialized structures such as cilia, flagella, contractile
filaments, muscle fibers or the mitotic apparatus. Although most
of the biochemical and physical details of motility are unexplained,
some of the more fundamental details have become classified primarily
through work on striated muscle systems.

In multicellular organisms, the functional contractile system
is referred to as a "motor unit". The motor unit consists of
receptor and neurological cells which make synaptic contact with
an effector, or muscle system. The receptor cell has the capacity
to detect a stimulus and elicit a specific contractile response to
that stimulus through chemcial neurotransmitters.

II. CONTRACTILITY IN MUSCLE CELLS

The functional contractile unit in vertebrate skeletal muscle
is the sarcomere. In series, the sarcomeres form a fibril which
makes tendonous connections with the skeletal system. Simultaneous
contractions of sarcomeres in series in a fibril result in movement
of the skeletal system at specific hinges. This coordinated action
results in locomotion. The outer membrane of a sarcomere, the
sarcolemma, makes synaptic contact with an innervating excitatory
cell at specific points along the membrane which have been described

as myoneural junctions. The excitatory stimulus from the innervating neuron is transferred chemically to the post-synaptic membrane
of the junction.

Specific enzyme receptors at this site respond to the release
of the transfer chemical by eliciting a depolarizing potential.
It has been suggested that the activating current is then spread
inwards by a system of membranes, the transverse (T) tubular
system, which are infoldings of the sarcolemma (Podolsky, 1971;
Eisenberg and Gage, 1969; Eisenberg, 1971; Huxley and Taylor,
1958). Ultrastructural observations of frog striated muscle fibers
by Peachey (1965) showed the lumen of the T-system to be continuous
with the extracellular space and occurred at regular intervals at
the borders of adjoining sarcomeres. This region of T-tubules is
referred to as the Z-line. A second system of membrane limited
spaces occurs lateral to the T-tubules and forms a network across
each sarcomere, from Z-line to Z-line. The spaces enclosed by
this system of membranes, the sarcoplasmic reticulum (S.R.), are
discontinous with the extracellular spaces described by the T-system
of membranes according to observations made by Huxley (1964) on the
sartorius muscle of frogs. The sarcoplasmic reticular system has
been shown to act as a calcium store by sequestering calcium from
the sarcoplasm through an energy-dependent, membrane-associated
process (Hasselbach and Makinose, 1961; Costantin, Franzini-Armstrong
and Podolsky, 1965; Ebashi, 1961). Current models for the activation of contraction suggest that the surface-membrane action potential is conducted inwards by way of the membranes of the T-system.
Presumably, through a process yet unknown, the inward flow of
current triggers a calcium reversal potential at the terminal sacks
of the sarcoplasmic reticular membrane. Calcium movement into the
sarcoplasm then initiates contraction.

III. CALCIUM ACTIVATION OF CONTRACTION

Thin filaments extend from each Z-line towards the center of
the sarcomere through the region referred to as the I-band. Past
the I-bands the thin filaments interdigitate with centrally located
thicker filaments which occupy the central region of the sarcomere,
the A-band. The thin filaments have been shown by Endo et al.
(1966) to contain filamentous actin, troponin and tropomyosin.
The thick filaments contain myosin arranged symmetrically on each
side of the median of the longitudinal axis of the sarcomere. The
boundaries of the A-band are determined by the lengths of the myosin
filaments. In biochemical preparations of actin and myosin, the
monomeric units of the two molecules can be induced to form actomyosin aggregates in the presence of magnesium ions and ATP (Maruyama
and Gergely, 1962; Tawada and Oosawa, 1969; Weber and Winicur, 1961).
In the presence of troponin and tropomyosin the actomyosin complex

dissociates. Interaction resulting in aggregation can only be
reestablished if calcium is added in concentrations above 0.1μM
(for review see A. Weber and R.O. Bremel, 1971). In summary,
contractility in striated muscle is initiated by the electrogenic
movement of calcium ions from storage and binding sites on the
membranes of the sarcoplasmic reticulum. The initial movement of
the ions is triggered by the movement inwards, along the membranes
of the T-tubules, of a propagated, Na^+, K^+-dependent action
potential. The availability of free calcium ions in the sarcoplasm
in concentrations approaching 1μM results in an interaction with
the regulator protein troponin. Conformational changes in troponin-
tropomyosin induced by calcium association results in the removal
of the allosteric inhibition of the interaction between actin and
myosin. Cyclical interactions between actin and myosin result
in the hydrolysis of ATP through a Mg^{++}-ATPase contained in the
myosin molecule and a shortening of the myofibril. Relaxation-
reextension molecule is achieved when the concentration of calcium
in the sarcoplasm is restored to a level of 0.1μM through the
process of active sequestration by the longitudinal elements of the
sarcoplasmic reticulum. This type of calcium-coupled contractile
system is found primarily in fast-acting mammalian striated (twitch)
muscle system.

IV. NON-MUSCULAR SYSTEMS AS MODELS FOR VARIOUS ASPECTS OF CONTRACTION
IN MUSCLE SYSTEMS

Attempts to carry out in vivo pharmacological analyses of the
events, in fast-acting muscle systems, which lead to contractility
are seriously hampered by problems of intercellular diffusion. These
problems are due primarily to the multicellular organization of
striated muscle systems. The stacking and alignment of sarcomeres
must certainly produce ionic and metabolic gradients with concomitant electrical gradients from the surface of muscle preparations
to the central axis. The direct effect of the existence of these
gradients is to impede accurate measurements of rapid changes in
metabolic activity throughout the system in response to chemopharmacodynamic treatment or other changes in environment. The most
difficult aspects of these problems can be bypassed through the use
of single-celled or protozoan cultures which have the same properties
of irritability or excitability as do muscle systems. Irritability
or excitability in such systems could be a locomotory or contractile
response to a local chemical, electrical, thermal, photic or
mechanical stimulus.

In order that the behavior of the systems, in response to the
type of stimuli mentioned above, be used to interpolate similar
behavior in muscle systems, an accurate documentation of the
differences and similarities between protozoan and muscle systems

is being undertaken. These efforts may be divided into three
major headings: (a) receptor mechanisms, specific and/or nonspecific,
which are membrane associated (see David Wood in this volume);
(b) excitation-contraction coupling mechanisms which are calcium-
dependent, and (c) the contractile architecture and biochemical
properties of the contractile proteins.

Ciliated protozoa of the sub-orders Heterotrichida and
Sessilinida posses the ability to perform very fast contractions
on the order of 4 msec. (Jones, Jahn and Fonseca, 1966, 1969).
It has been suggested that the contractile behavior of these
organisms is a protective response to possible damage to their
feeding apparatus; or, for burrowing in interstitial sediment
(Pautard, 1960; Dragesco, 1962). One further possibility is that
the contractile activity induces local turbulence in the medium
causing rapid diffusion of toxic by-products away from the organism
and an increase in useful metabolites and oxygen. Contraction
may be 80% to 90% of the original length, as observed by Randall
and Hopkins (1962) in Carchesium. In Spirostomum (Figure 1)
and Stentor, shortening may reach a maximum of 45% of their original
lengths (Jones et al., 1966; Ettienne, 1970; Newman, 1972; see also
Hamilton in this volume [Ed.]). Contractility in Stentor and
Spirostomum is directed by longitudinally arrayed fibrillar systems
adjacent to the cell cortex (Figure 2a, b). Initial observations
by Randall and Jackson (1958), Grain (1968), and Bannister and
Tatchell (1968) resulted in the identification of distal longitudinal ribbons of microtubules within membrane folds, and more
proximally, longitudinally arrayed, bands of microfilaments of
40-50 Å diameter. Similar observations were made on Spirostomum
by Grain (1968), Ettienne (1970) and Lehman and Rebhun (1971).
By convention, the overlapping microtubular ribbons are termed
kinetodesmal fibers (Km) and the microfilamentous bundles, myonemal
(M) fibers (Pitelka, 1969). Observations by Lehman and Rebhun (1971)
and Newman (1972) showed that the myonemal fibers in Spirostomum
and Stentor change in distribution and appearance so as to suggest
that they are undergoing an active shortening during the development of contractile tension. Similar observations by Huang and
Pitelka (1971) on Stentor substantiate the reported deformation
of the myonemal fibers and further suggest that the Km fibers
telescope by each other and, that the normal free-swimming state
of the organism is reestablished by the active interaction of cross-
bridges on adjacent tubules of the Km fibers. This interaction,
first proposed by McIntosh (1971), appears to be inhibited by an
increase in intracellular free calcium and is dependent on metabolic
energy (Huang and Pitelka, 1971; Ettienne, 1974). Studies on
glycerinated models of Stentor, Vorticella and Spirostomum, show
that myonemal contraction does not seem to be dependent on the
hydrolysis of ATP. Contraction can be elicited solely by an increase
in calcium concentration above 0.1µM, approaching the levels of

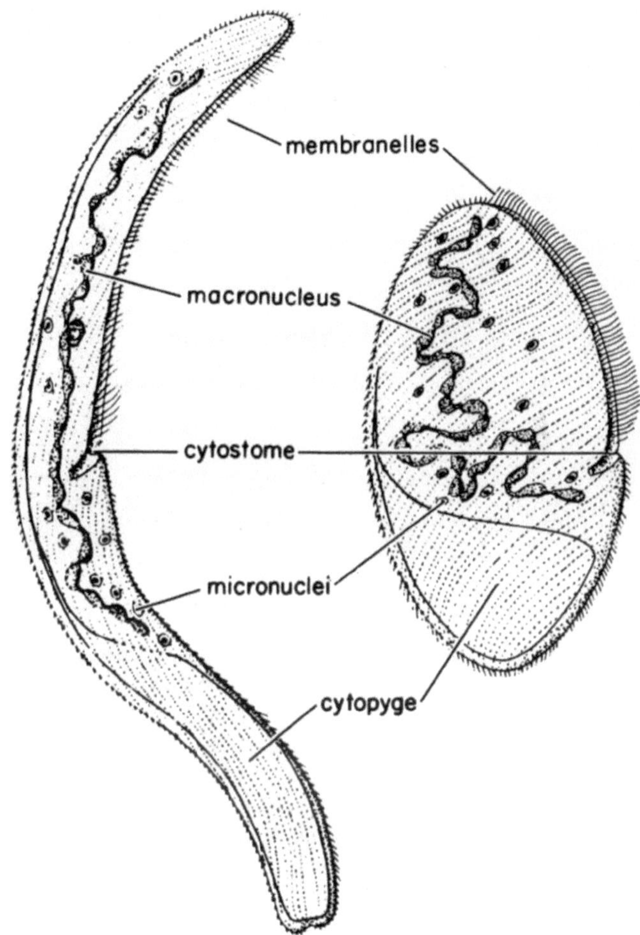

Figure 1. Illustration of the free-swimming and contracted states
of the heterotrichous ciliate, _Spirostomum_ _ambiguum_. Adult
organisms of this species are cylindrical in shape and may have
dimensions of 3 x 0.3 mm. Contractions to as much as 50% of
length can be spontaneous or in response to electrical, chemical,
mechanical or photic stimulation.

calcium necessary to initiate contraction in muscle cells. The
mechanism of contraction of this calcium-activated, energy independent process remains to be elucidated. However, Weis-Fogh and
Amos (1972) suggest that the contractile myonemes are rubber-like
polymers with repulsive ionic charges which can be neutralized
through the binding of divalent cations. Cationic binding leads
to a conformational change in the polymer which results in shortening. It is interesting to note that while calcium induced contraction

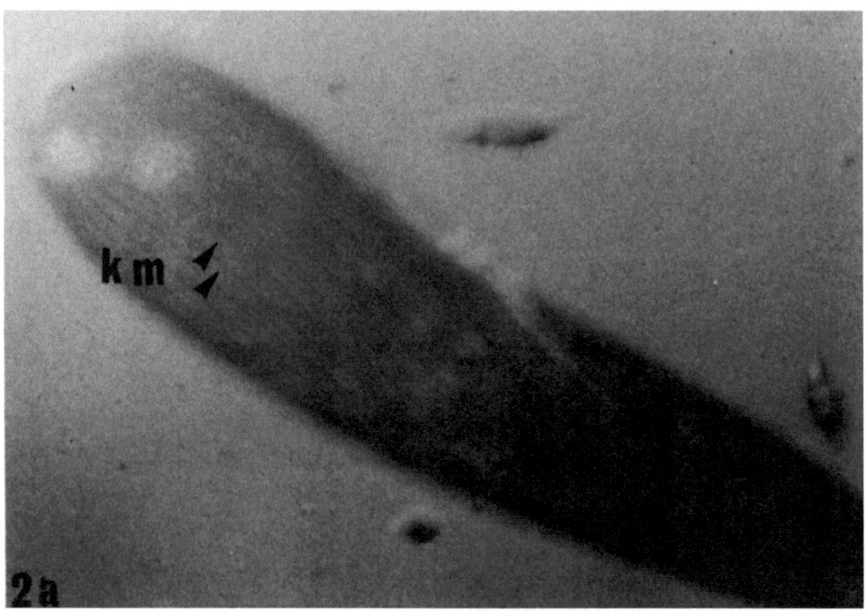

Figure 2a. Optical image of a living _Spirostomum_ viewed through a
Zeiss polarization microscope. The negatively birefringent bands
running at a slight angle to the longitudinal axis of the organism
are bundles of microtubules or Kinetodesmal (Km) fibers.

in glycerinated models can be reversed cyclically by alternately
lowering and raising the free calcium ion concentration around
0.1µM in the presence of classical metabolic antagonists, in living
Spirostomum treated under similar conditions, contraction is sus-
tained and irreversible and the Km fibers are destroyed. However,
brief tetanic contractions are superimposed, under these conditions,
on the contracted organism by electrical stimulation. The pharmacological agent, cytochalasin, which acts as a potent inhibitor
of cell division with a reputed effect on contractile filaments,
(a) blocks contractile activity in _Spirostomum_ in response to all
stimulation (including fixatives); (b) induces asynchronous
ciliary activity, and (c) causes an elongation of the cell to
125% of its original length without any obvious changes in micro-
filament or microtubular ultrastructure (Figure 3) (Ettienne, 1974).
Information on the site of action of cytochalasin is still very
limited. However, results suggest that in _Spirostomum_, it may act
on membrane-myonemal linkage complexes. Such complexes have been
suggested for the attachment of the _Vorticella_ "spasmoneme" to the
cell membrane by way of basal body-like structures (Allen, 1973).
Basal bodies also anchor cilia to the membrane of _Spirostomum_.

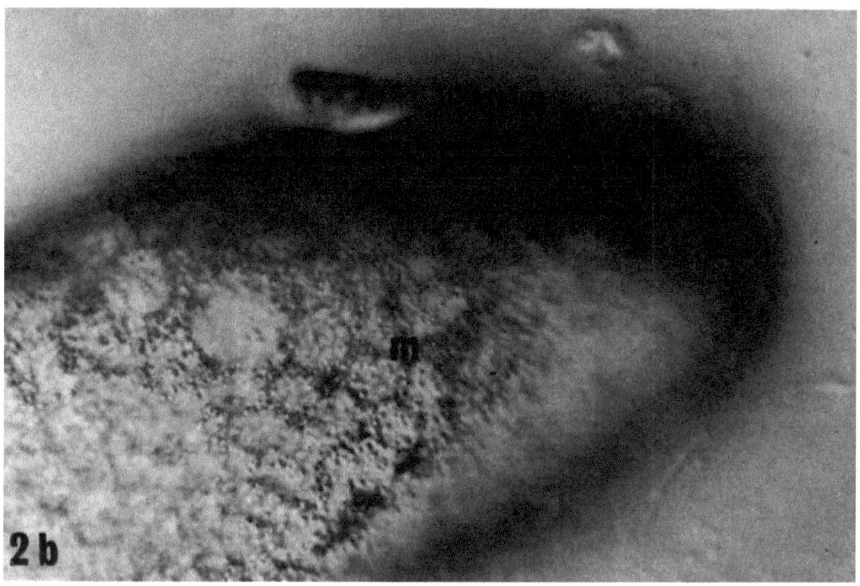

Figure 2b. Photomicrography of the myonemal (M) network of
Spirostomum which is outlined by the dense aggregation of mitochondria. The myonemal filaments in aggregate form pentagonal lattices
which extend around the organism on a plane approximately 8-10μM
below the pellicle.

A "disruption" of ciliary basal bodies by cytochalasin could also
explain the observed asynchronicity and loss of cilia from the
cell following treatment with cytochalasin.

     If one accepts that the Km microtubules and the myonemal
filaments act antagonistically (Huang and Pitelka, 1971), and that
the contractile filaments have elastic properties similar to rubber
(Weis-Fogh and Amos, 1972), then an increase in the elastic exten-
sibility (or detachment from fixed points on the membrane) of the
filaments could explain the observed extension of the cell due to
the antagonistic, tensile action of the microtubules following
treatment with cytochalasin.

          Calcium Localization and Calcium Release in Spirostomum

     It is evident from studies on muscle and other related con-
tractile systems that the contractile response fails when the
calcium concentration falls below 0.1μM (Weber, 1966; Weis-Fogh
and Amos, 1972; Huang and Pitelka, 1971). Initial speculation on

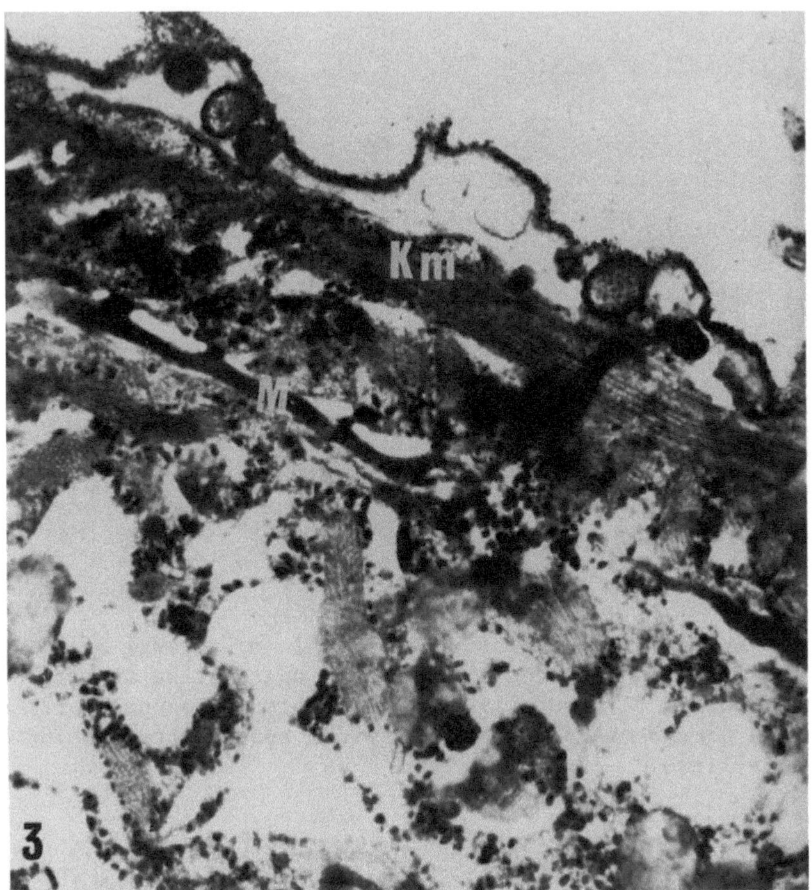

Figure 3.   Cross-section of <u>Spirostomum</u> treated with 1 mg/ml.
cytochalasin and fixed in 2% glutaraldehyde-acrolein and post-
treated with 2.5% osmium tetroxide.   The myonemal (M) and kinetodesmal (Km) fibers are preserved apparently in their natural
states in the extended organism following treatment with cytochalasin.
Apparently cytochalasin blocks contraction normally induced by
fixatives. x 12,000.

the role of calcium ions as initiators of contraction was provided
by Heilbrunn and Wiercinski (1947) who showed that calcium alone,
among a number of cations injected into striated muscle fibers,
initiated contraction.   Heilbrunn's pioneering work was later confirmed by the experiments of Niedergerke (1955) who electrophoreti-
cally injected calcium into the muscle fibers of the spider crab.
Subsequent investigations on vertebrate striated muscle confirmed

the release of calcium from internal stores following electrical
excitation and its subsequent removal, leading to relaxation (for
review see J. Podolsky, 1971).

When _Spirostomum_ are treated with 10mM sodium oxalate, according to the technique of Costantin _et al_. (1965), there is a progressive and marked decrease in contractile response to external
stimuli. Electron micrographs of _Spirostomum_ taken from fresh
cultures and pre-treated with oxalate before fixation, often
revealed multiple electron-dense precipitates within membrane-limited
spaces. Ten micron resolution electron microprobe-microanalysis
indicated that the precipitates were abundant in calcium (Ettienne,
1970). It was also determined that organisms microinjected with
calcium alone contracted as opposed to the microinjection of
magnesium, sodium, and potassium ions (Ettienne, 1970; Ettienne
and Selitsky, 1974).

In order to show that calcium release from internal stores
was indeed a phenomenon which accompanied contraction in the organism, 0.2 u L aliquots of aequorin (o.143 mg/ml), supplied by Dr.
O. Shimomura of Princeton University, were injected into organisms.
Aequorin is a photoprotein from the jelly fish _Aequorea_ which emits
light only in the presence of calcium ions (Shimomura _et al._, 1962,
1963a; Hastings _et al._, 1969). Light emission is proportional to
the aequorin concentration over a wide range of calcium ion concentrations. Thus, aequorin was used primarily to measure concentration changes for cytoplasmic free calcium in _Spirostomum_ (Ettienne,
1970). Figure 4 shows sample traces of the photomultiplier output
(RCA 8575) when _Spirostomum_ were injected with aequorin and allowed
to sit on the surface of the phototube in a transparent chamber
for extended time periods and in the absence of direct external
stimulation. The traces show occasional pulses which may have
resulted from spontaneous contractions of the organism. The light
output is well above the baseline activity or dark current of the
phototube and within the range of sensitivity of the human eye.

Electrically induced contraction occurred with a mean latency
of 20.5 msec in organisms injected with aequorin, and had a duration greater than 500 msec for the stimulus strength and duration
used. When traces of the photomultiplier output were superimposed
on the shortening and relaxation cycle for the same organism as
determined by 400 f.p.s. cinéfilm analysis, it was clear that
shortening did not begin until the calcium-aequorin light emission
was at a peak. Relaxation begins when light emission drops off,
implying a removal of calcium ions to other membrane-limited
cellular compartments (Figure 5).

Measurement of the electrical properties of the membrane
showed that the organism has a mean resting membrane potential of
15 mV, negative inside. The mean value of the specific membrane

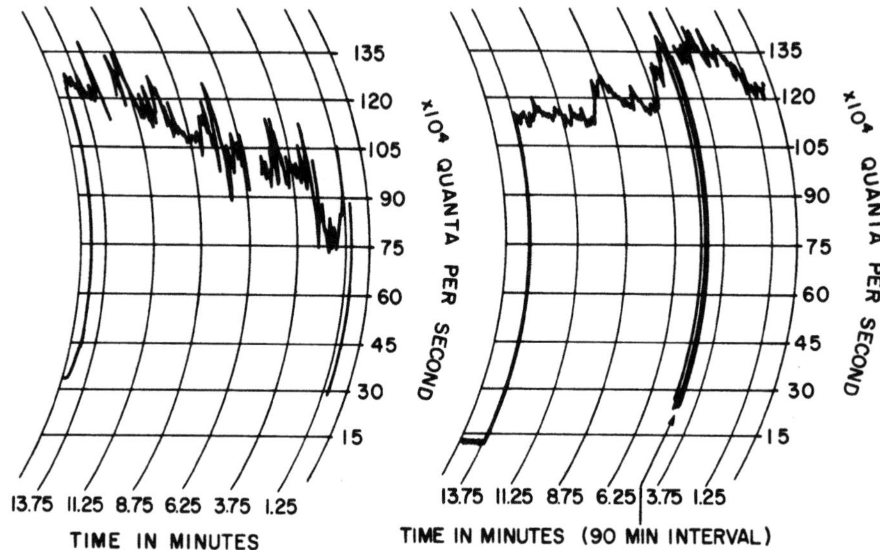

Figure 4. Chart recordings of photomultiplier (RCA 8575) current
changes induced by light output from _Spirostomum_ microinjected with
0.143 mg/ml of the bioluminescent protein aequorin. Occasional
spontaneous photoemissions from the cell may be greater than 1.4 x
10^6 photons per second, well within the sensitivity of the human
eye. The enzymic degradation and decay of aequorin in the organism
is a very slow process as evidenced by the fact that light output
has decayed only by about 12% after more than 2 hours in the living
organism (Chart B).

resistance was $(25 \pm 5.85$ SD) x 10^4 ohm. cm^2, more than twice that of
muscle cells. There was no measurable depolarization of the cell
membrane following the application of electrical stimuli greater
than threshold for mechanical response. Thus, it is assumable that
the reception and transmission of a signal to contract operates in
a manner different from that of muscle. Perhaps reception of the
signal is a passive, tonic effect with very small depolarizations
(2-3mV). It may involve the release of calcium ions from membrane-
binding sites within the cell with no large changes in charge distribution across the membrane. This concept is somewhat substantiated by the lack of measurable depolarization and the fact that
work in progress shows that the classical muscle cholinergic agonists
and antagonists have very little effect on contractitlty in
Spirostomum and the related ciliate, _Vorticella_, at "physiologically
acceptable" concentrations (Maran _et al._, 1972). However, contraction is inhibited by calcium ionophores, some barbituates,
reducing agents, and metabolic poisons (author's own observations).
Obviously, the problem of identification of effector-receptor
mechanisms is a very important one for this organism. Perhaps the

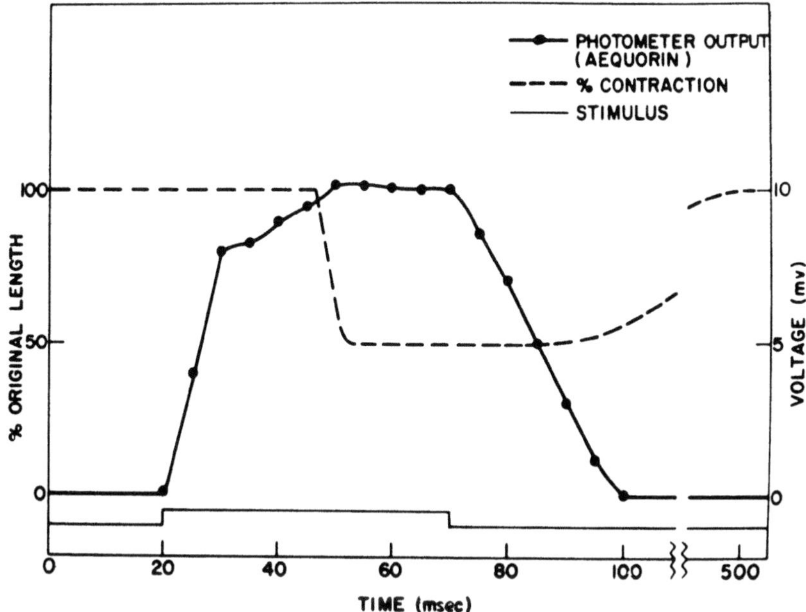

Figure 5. Graph shows the relationship between the onset of electrical stimulation (——), photon emission (-●-), and maximum
latency to contraction (----). Mechanical contraction begins when
aequorin-calcium induced photon emission is at a peak. Photon
emission drops off and relaxation ensues with the cessation of
electrical stimulation.

approach we are using in looking for specific receptors such as
exist in muscle is too narrow. Since the organism responds to such
a wide variety of environmental stimuli, it would seem more probable
that it has a much less specific, non-linear, receptor system which
is potentiated by stimulation. Since the coupling of reception to
contraction is apparently governed by calcium, it would seem more
likely that substances which affect calcium binding or accumulation
by membranes would be effective agents with which to begin a study
of such a general "receptor" mechanism. Thus, it seems that the
only true information we have on Spirostomum are the following:

(1) The organism responds to chemical, electrical, mechanical,
    thermal, and photic stimuli, either by avoidance through
    locomotory behavior or contraction.

(2) Calcium appears to be an indirect coupling agent for
    excitation-to-contraction.

(3) There are elements within the cell which are arranged in
    such a manner as to exert either tensile or contractile
    forces on the organism.

(4) The elements responsible for the generation of a tensile
    force appear to be the microtubular ribbons which are the
    principal components of the Km fibers. The interactions
    of these elements in <u>Stentor</u> appear to be inhibited by
    increases in cytoplasmic free calcium.

(5) The contractile force is generated by the myonemal (M)
    bands which are possibly rubber-like, 40-50 Å-diameter
    filaments which are calcium activated. We know very
    little about the biochemistry of the contractile proteins
    or the nature of the receptor mechanism (S).

(6) Contraction in <u>Spirostomum</u> is similar to contraction in
    muscle in that they may both respond to certain kinds of
    stimulation and in both systems calcium acts as a final
    initiator of contraction.

V. ROLE OF CALCIUM IN CYCLICAL DEFORMATIONS OF <u>PHYSARUM POLYCEPHALUM</u>

A different class of organisms whose locomotory activity is
based on cytoplasmic streaming have been shown to possess proteins
with physical and chemical properties similar to those of muscle
actin and myosin (Hatano <u>et al</u>., 1966a, 1966b, 1968; Adelman and
Taylor, 1969a,b; Pollard and Ito, 1970; Pollard <u>et al</u>., 1970).
Ultrastructural fibrils consisting of densely packed microfilaments
similar in size to the thin filaments of striated muscle have been
observed in the plasmodium of the acellular slime mold, <u>Physarum
polycephalum</u>. These filaments are often seen in association with
membrane-limited vesicles within organisms fixed for electron
microscopic observation (Rhea, 1966; Wohlfarth-Bottermann, 1964).

Pre-treatment of the organism with 10 mM sodium oxalate prior
to fixation resulted in the massive deposition of precipitates within
the region of the advancing margin of the plasmodium (Figures 6 and
7). Streaming is markedly slowed by treatment with the oxalate
solution, and there is retraction and detachment of the advancing
margin from the substratum. Vesicles containing election opaque
deposits were present only in sections taken of organisms treated
with oxalate.

Microprobe microanalysis of the oxalate induced precipitates
indicated a relatively high abundance of calcium with a strong
primary (Kα) x-ray emission and two secondary (Kβ) peaks (Figure 8).
The only other element present in concentrations above noise level
was potassium (Ettienne, 1972).

The localization of calcium stores in <u>Physarum</u> suggests a mechanism for control of the cyclical deformations of the plasmodium

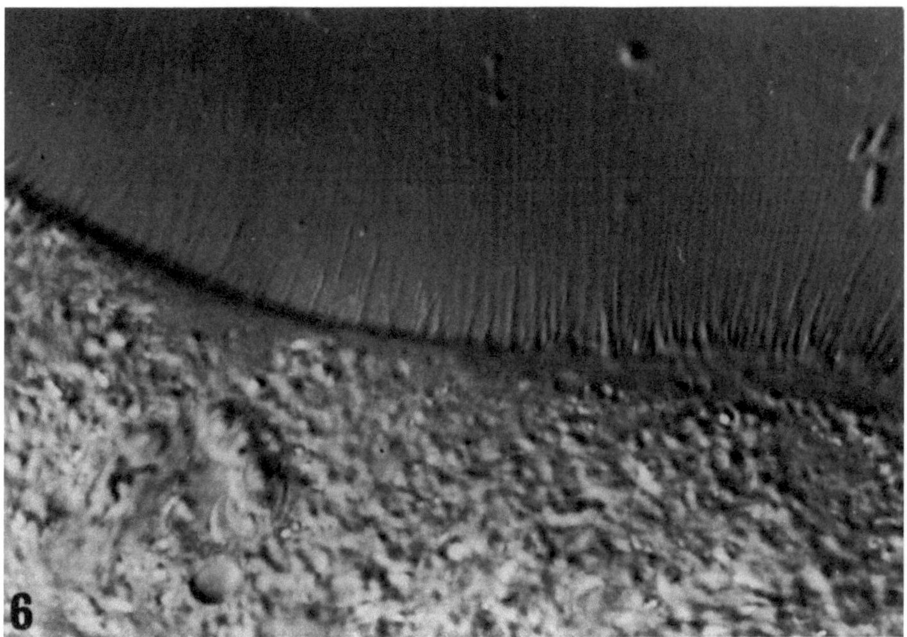

Figure 6. Zeiss-Nomarski differential interference photomicrograph
of the advancing margin of the slime mold, Physarum polycephalum.
The region is populated by numerous rotating nuclei (lower left),
and microvilli-like projections to the agar substrate. x 1500.

through a calcium release and sequestration mechanism as has been
described for muscle and other contractile cells. The establishment
of a biochemical identity for the contractile proteins which appears
to be like that of striated muscle protein lends further credence
to a calcium activation mechanism. The vesicles could ostensibly
control the levels of free calcium available to the actomyosin-
like ATPase system.

The possibility of calcium release through a depolarizing
potential is substantiated by the existence of a cycle of bioelectri-
cal potential change superimposable on the cyclical pattern of
plasmodial deformations and cytoplasmic streaming, with a slight
phase shift (Kamiya and Abe, 1950; Tauc, 1953). Because the potential changes persist when streaming is stopped by applied physical
force equal and opposite to the hydraulic pressure gradient in the
streaming plasmodium, it is assumed that the electrical changes

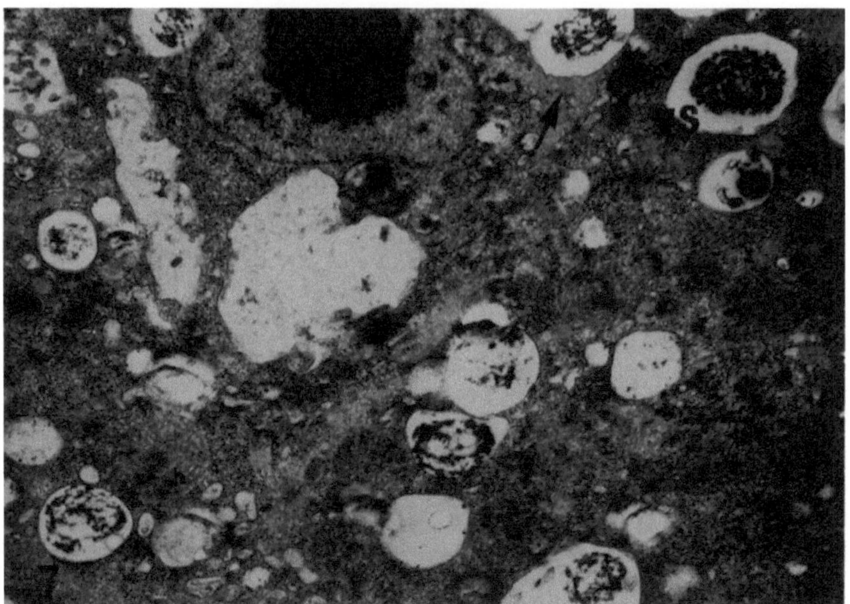

Figure 7. Section of the plasmodium of _Physarum_ treated with 10mM
sodium oxalate and fixed in 2% glutaraldehyde. Oxalate precipitates
are seen in vesicles (Vs) which appear throughout the plasmodium.
Arrow points to a lightly staining area which shows the attachment
of microfilaments to the vesicular membrane. X 10,000.

are not a result of electrical fields generated during streaming,
but rather an independent event related to the motive force for
streaming. These electrical potential changes might possibly serve
as triggers for calcium release and thus regulate the levels of
calcium ions in different regions of the plasmodium. Because of
its wide range of specific responsivity to stimuli, particularly
photic (Daniel and Jarlfors, 1972), and also because of its large
yield in a variety of culture conditions, _Physarum_ appears particularly amenable to the study of receptor-effector mechanisms.
It is in fact a dark-horse candidate which could yield very
valuable information on a wide range of cellular phonomena.

The field is wide open for research on the types of "primitive"
mobile and contractile phonomena discussed in this chapter. The
watergates need only be opened.

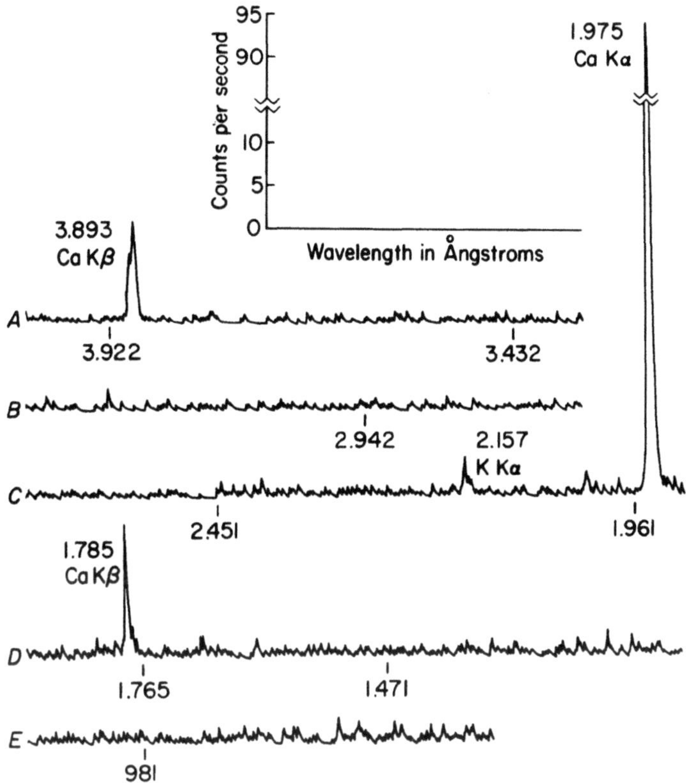

Figure 8. Continuous spectral readout (lines A through E) from a thick, frozen section of a plasmodium treated with 10 mM sodium oxalate. The analysis of the precipitates was accomplished through the use of a microprobe-microanalyzer (CAMECA). Lines A and D show two secondary emission peaks (Kb) for calcium X-rays. Line C shows the primary (Ka) emission peak for calcium as well as a weak (Ka) emission for potassium.

REFERENCES

Adleman, M.R. and Taylor, E. 1969a. Isolation of an actomyosin-like protein complex from slime mold plasmodium and the separation of the complex into actin and myosin-like fractions. Biochem., 8, 4964.

Adelman, M.R. and Taylor, E. 1969b. Further purification and characterization of slime mold myosin and slime mold actin. Biochem., 8, 4976.

Allen, R.D. 1973. Structures linking the myoneme, endoplasmic reticulum and surface membranes in the contractile ciliate Vorticella. J. Cell. Biol., 56, 559-579.

Bannister, L.H. and Tatchell, E.C. 1968. Contractility and the fibre systems of Stentor Coeruleus. J. Cell. Sci., 3, 295.

Costantin, L.L., Franzini-Armstrong, C. and Podolsky, R.J. 1965.
  Localization of calcium-accumulating structures in striated muscle
  fibers. Science, 147, 158.
Daniel, J.W. and Jarlfors, V. 1972. Light-induced changes in the
  ultrastructure of a plasmodial myxomycete. Tissue & Cell, 4 (3),
  405-426.
Dragesco, J. 1962. On the biology of sand-dwelling ciliates.
  Sci. Prog. (London), 50, 353-363.
Ebashi, S. 1961. Calcium binding activity of vesicular relaxing
  factor. J. Biochem. Tokyo, 50, 236.
Eisenberg, R.S. 1971. The equivalent circuit of frog skeletal
  muscle fibers. In Contractitlty of Muscle Cells and Related
  Processes. pp. 69-88.
Eisenberg, R.S. and Gage, P.W. 1969. Ionic conductances of the
  surface and transverse tubular membranes of frog fibers. J. Gen.
  Physiol., 53, 279.
Endo, M., Nomomura, J., Mosaki, T., Ohtusuki, I. and Ebashi, S.
  1966. Localization of native tropomyosin in relation to striation
  patterns. J. Biochem., 60, 605.
Ettienne, E.M. 1970. Calcium Regulation of Contraction in non-
  muscular contractile systems. Doctoral Thesis, S.U.N.Y., Albany.
Ettienne, E.M. 1970. Control of Contractility of Spirostomum by
  dissociated calcium ion. J. Gen. Phys., 56, 168.
Ettienne, E.M. 1972. Subcellular localization of calcium reposi-
  tories in plasmodia of the acellular slime mold, Physarum
  polycephalum. J. Cell. Biol., 54, 179-184.
Ettienne, E.M. and Selitsky, M. 1974. The antagonistic effects of
  antimitotic agents on contraction and relaxation in Spirostomum
  Ambiguum. J. Cell Science, Nov. (In Press).
Grain, J. 1968. Les systemes fibrillaires Chez Stentor igneus
  Ehrenberg et Spirostomum Ambiguum Ehrenberg. Protistologica, 4,
  27.
Hasselbach, W. and Makinose, M. 1961. Die Calciumpumpe der
  "Erschlaffungsgrana" des Muskels und ihre abhangigkeit von der
  ATP spaltung. Biochem. Z., 333, 518.
Hastings, J.W., Mitchell, G., Mattingly, P., Blinks, J. and Van
  Lecuwen, M. 1969. Response of aequorin luminescence to rapid
  changes in calcium concentration. Nature (London), 222, 1047.
Hatano, S. and Oozawa, F. 1966a. Extraction of actin-like protein
  from the plasmodium of a myxomycete and its interaction with
  myosin A from rabbit striated muscle. J. Cell. Physiol., 68, 197.
Hatano, S. and Oozawa, F. 1966b. Isolation and characterization
  of plasmodium actin. Biochem. Biophys. Acta., 127, 488.
Hatano, S. and Tazawa. 1968. Isolation, purification and characterization of myosin B from myxomycete plasmodium. Biochem.
  Biophys. Acta., 154, 507.
Huang, B. and Pitelka, D.R. 1971. The Contractile Process in the
  Ciliate Stentor Coeruleus, I. Functional role of Microtubules
  and Microfilaments. Doctoral Dissertation, U.C., Berkeley.

Huxley, H.E. 1964. Evidence for continuity between the central elements of the triad and extracellular space in frog sartorius muscle. Nature, 202, 1067.

Huxley, A.F. and Taylor, R.E. 1958. Local activation of striated muscle fibres. J. Physiol. (London), 144, 426.

Jones, A.R., Jahn, T.L. and Fonseca, J. 1966. Contraction of protoplasm. I. Cinematographic analysis of the anodally stimulated contraction of Spirostomum Ambiguum. J. Cell. Phys., 68, 127-134.

Jones, A.R., Jahn, T.L. and Fonseca, J. 1970. Contraction of protoplasm. IV. Cinematographic analysis of the contraction of some peritrichs. J. Cell Physiol., 75, 9-20.

Kamiya, N. and Abe, S. 1950. Bioelectric phenomena in the myxomycete plasmodium and their relation to protoplasmic flow. J. Colloid. Sci., 5, 149.

Kandel, Eric. 1970. Nerve cells and behavior. Sci. Amer., pp. 57-70.

Lehman, W.J. and Rebhum, L.J. 1971. The structural elements responsible for contraction in the ciliate Spirostomum. Protoplasma, 72, 153.

Maran, M., Himmelstein, R. and Dikstein, S. 1972. Vorticella -- a model for chemopharmacodynamic action on smooth muscle. Comp. Gen. Pharm., 3 (11), 363-370.

Maruyama, K. and Gergely, J. 1962. Interaction of actomyosin with adenosine triphosphate at low ionic strength. II factors influencing clearing and superprecipitation: adenosine triphosphatase and birefringence of flow studies. J. Biol. Chem., 237, 1100.

McIntosh, J.R. 1971. Microtubule contraction and sliding associated with cellular motility. Abst. 11th Ann. Meet. Amer. Soc. Cell. Biol.

Newman, E. 1972. Contraction in Stentor coeruleus: a cinematic analysis. Science, 177, 447.

Niedergerke, R. 1955. Local muscular shortening by intracellularly applied calcium. J. Physiol. (London), 128, 12.

Pautard, F.G. 1960. Calcification in unicellular organisms. In Calcification in Biological Systems, R.F. Sognnaes (Ed.), AMS Washington.

Peachey, L.D. 1965. The sarcoplasmic reticulum and transverse tubules of the frog's sartorius. J. Cell. Biol., 25, 209.

Pepe, F.A. 1966. Some aspects of the structural organization of the myofibril as revealed by antibody-staining methods. J. Cell. Biol., 28, 505.

Pitelka, D.R. 1969. Fibrillar systems in protozoa. In Research in Protozoology, 3, 280, T.T. Chen (Ed.), Pergamon Press, Oxford and N.Y.

Podolsky, R.J. 1971. Contractility of Muscle Cells and Related Processes, Prentice Hall (New Jersey).

Pollard, T. and Ito, S. 1970. Cytoplasmic filaments of amoeba proteus. I. Role of filaments in consistency changes and movement. J. Cell. Biol., 46, 267.

Pollard, T., Shelton, E., Witting, R. and Kora, E.D. 1970.
   Ultrastructural characterization of F-actin isolated from
   Acanthamoeba castellanii and identification of cytoplasmic
   filaments as F-actin by reaction with rabbit heavy meromyosin.
   J. Mol. Biol., 50, 91.
Randall, J.T. and Hopkins, J.M. 1962. On the stalks of certain
   peritrichs. Phil. Trans. Roy. Soc. (London) B, 245, 59.
Randall, J.T. and Jackson, S.F. 1958. Fine structure and function
   in Stentor polymorphus. J. Biophys. Biochem. Cytol., 4, 807.
Rhea, P.R. 1966. Electron microscopic observation on the slime
   mold, Physarum polycephalum with specific reference to fibrillar
   structures. J. Ultrastruct. Res., 15, 349.
Shimomura, O., Johnson, F. and Saiga, J. 1962. Extraction purification and properties of aequorin, a bioluminescent protein
   from the luminous hydromedusa, Aequorea. J. Cell. Comp. Physiol.,
   59, 223.
Shimomura, O., Johnson, F. and Saiga, Y. 1963a. Microdetermination
   of calcium by acquorin luminescence. Science, 140, 1339.
Tauc, L. 1953. Quelques observations de bioélectricité cellulaire,
   en particulier chez un myxomycete. (Physarum polycephalum).
   J. Cell. Biol., 54, 179-184.
Tawada, K. and Oozawa, F. 1969. Activation of H-meromyosin ATPase
   by polymers of actin and carboxymethylated actin. J. Mol. Biol.,
   44, 309.
Weber, A. and Bremel, R. 1971. Regulation of Contraction and
   Relaxation in the myofibril. In Contractility of Muscle Cells and
   Related Processes, R.J. Podolsky (Ed.), Prentice Hall, pp. 37-53.
Weber, A. and Winicur, S. 1961. The role of calcium in the super-
   precipitation of actomyosin. J. Biol. Chem., 236, 3198.
Weis-Fogh, T. and Amos, W.B. 1972. Evidence for a new mechanism
   of cell motility. Nature, 236, 301-304.
Wohlfarth-Bottermann, K.E. 1964. Differentiations of the ground
   cytoplasm and their significance for the generation of the motive
   force of amoeboid movement. In Primitive Motile Systems in Cell
   Biology, R.D. Allen and N. Kamiya (Eds.), Academic Press, New
   York, p. 79.

BEHAVIORAL PLASTICITY IN PROTOZOANS

Thomas C. Hamilton*
Department of Zoology
University of Texas at Austin
Austin, Texas 78723

INTRODUCTION

It is perhaps obvious that investigations of aneural systems
(particularly protozoans) can provide insight into problems concerning receptor-transducer mechanisms as well as the cellular
control of both ciliary and contractile systems. For the biochemical
homology which pervades the animal kingdom suggests that the neurons,
receptor cells, muscle fibers, etc. of metazoans use molecular mechanisms similar to those employed by protozoans. In fact the
generalized nature of unicellular organisms which, within the confines of one cell, contain the regulatory mechanisms necessary for
the survival of an entire organism, further suggests that most
protozoans will have cellular mechanisms similar to a wide variety
of metazoan cell types.

It is perhaps less obvious, however, that these same organisms
possess a behavioral repertoire capable of meaningful correlation
with the adaptive phenomena normally associated with neurons and
neural systems. To those unfamiliar with the behavioral capabilities of protozoans, who appreciate these animals only as minute,
single-celled organisms with very limited and stereotyped behavioral
capabilities, the suggestion that behavioral investigations of protozoans can shed light upon the complex workings of a nervous system
may seem absurd. However, protozoans do indeed possess a wide
variety of behaviors, which provide the possibility of numerous
behavioral experiments. Additionally, some of these behaviors
modify over time in response to repetitive stimulation in manners
that empirically resemble adaptive processes associated with neural
systems.

---

*Present address: Central Intelligence Agency, Washington, D.C. 20505.

Still, it is conceptually difficult to visualize how behavioral
studies of unicellular organisms can possibly contribute to the
understanding of behavioral control and behavioral modification in
multicellular animals. Indeed, if one tries to equate behavioral
evaluations obtained from an organism that happens to be a protozoan
to those behaviors resulting from the integrated output of highly
specialized nervous systems, the analogy is not obvious. But if
viewed from the perspective of separate receptor and effector (and
possibly transducer) processes, these single cells are not without
similarities to the sensory-motor distinctions made in multicellular
organisms. If one further considers the behavioral observations of
protozoa to be behavioral evaluations taken on single cells, rather
than single organisms, protozoa can then be used to study the manner
in which a single cell can control and modify its own output. Thus,
a single protozoan might be viewed as an isolated neuron, the output of which is independent of variables such as presynaptic input.
Changes in the motor output or contractile probability over time,
which are observed in many ciliated protozoans, may well resemble,
and indeed may even utilize molecular mechanisms homologous to
those used by neurons and muscle fibers which show facilitatory
processes or which antifacilitate (habituate). Furthermore, studies
of the ability of protozoans to exhibit learning and learning related
phenomena provide insight into whether multicellular organization
is necessary for the existence of such phenomena.

Thus, protozoans provide model systems in which the modifiable
aspects of cellular output can be studied free from intercellular
connectivities, with all morphological components responsible for
the initiation and coordination of behavioral output localized within the confines of one isolated and identifiable cell. Furthermore,
the chemical milieu of the environment in which these animals live
can be precisely controlled and regulated to suit experimental
conditions. Relative to most metazoan systems where cells are
bathed in an interstitial fluid, the composition of which is
influenced both passively and actively by the other cells of the
organism, and in which adaptive phenomena usually cannot be localized to a single cell or group of cells, protozoans certainly
provide a great experimental simplification.

BEHAVIORAL EXPERIMENTS IN PROTOZOA

As discussed by Corning and VonBurg (1973), in their extensive
review of protozoan learning, behavioral plasticity in protozoans
was reported about the turn of the century (Jennings, 1901, 1906).
In these studies Jennings discovered that _Stentor_, a contractile,
ciliated protozoan, became progressively less responsive (likely to
contract) to repeated touches with either a glass rod or a hair.
Subsequent studies have repeatedly demonstrated the presence of

similar response decrements in other protozoa. Thus, habituation,
defined as the waning of the amplitude or the frequency of occurrence
of a response over time to discrete and repetitive stimulation, is
commonly observed among protozoa.

In fact, habituation is common throughout the animal kingdom
(Harris, 1943). Thompson and Spencer (1966) have provided nine
parameters which can be used to describe habituation and to compare
experimental results from different systems. These parameters
summarize the more salient features of habituation and emphasize
the importance of the temporal aspects of stimulus presentation,
the occurrence of spontaneous recovery of responsiveness after
stimulation is withheld, and the usual presence of a phenomenon
called dishabituation.

Eisenstein and Peretz (1973) have made use of these parameters
in their recent review of habituation in invertebrate systems
(Table 1). In general, they found that three properties (response
decrement, spontaneous recovery and dishabituation) are common to
almost all habituating systems. As will be discussed shortly, only
protozoa differ from this generalization (Table 1) in that dishabituation has not been found in this phylum. However, many of
the other parameters suggest that habituation in protozoans closely
resembles neurally mediated habituation, and thus may result from
similar cellular mechanisms. Furthermore, a study of protozoan
habituation may help to identify those regulatory processes unique
to nervous tissue from those which are common to most living systems.

In this brief review, only the ciliated protozoans _Spirostomum_,
Stentor, and _Tetrahymena_ (Figure 1) will be considered. Like
Paramecium these ciliates have a normal, spiraling mode of forward
swimming and can reverse the direction of their ciliary beat and
swim backwards. Additionally, _Stentor_ and _Spirostomum_ possess
contractile systems capable of shortening the animals to less than
50% of their resting lengths in the space of only a few milliseconds.
Stentor, a funnel-shaped organism, has the further ability to attach
the tip of its funnel to the substrate via pseudopodal outgrowths
and remain sessile. Both ciliary reversals and contractions can
be initiated by several modalities of stimulation [e.g., electrical,
tactile, mechanical (vibratory), chemical, or light (Blättner, 1926].

Behavioral experiments on _Spirostomum_ and on _Stentor_ often have
been concerned with the occurrence of habituation of contraction
probability to repetitive mechanical stimulation. Figure 2 (Osborn
_et al._, 1973) shows typical results from experiments conducted on
Spirostomum ambiguum. Ten groups of four _Spirostomum_ each were
stimulated for ten minutes at a rate of one stimulus per ten seconds.
A rapid decrease in the frequency of contraction occurred over the
stimulation period. One minute of stimulation was then given after

TABLE 1

	Intact Mammal	Spinalized Cat	Intact Snail	Acute and Semi-Intact Aplysia	Aplysia Gill c/o CNS	Crayfish Nerve-Muscle	Isolated Ganglion of Roach	Coelenterate	Protozoa: Spirostomum	Stentor
1. Repeated stimulation of the same stimuli results in decreased response. Decrease is negative exponential function of the number of applications.	X	X	X	X	X	X	X	X	X	X
2. Spontaneous recovery-Response recovers overtime if stimulus withheld.	X	X	X	X	X	X	X	X	X	X
3. With repeated series of habituation training, spontaneous recovery between series, habituation becomes more rapid.	X	X	X	O	X	--	--	X	X	--
4. The more rapid the rate of stimulation, the more rapid the habituation.	X	X	--	X	X	--	X	--	X	X
5. Weaker stimulus elicits more rapid habituation.	X	X	--	X	--	--	--	--	X*	X
6. Effects of habituation training may proceed beyond zero or asymptotic level.	X	X	--	O	--	--	--	--	X*	--
7. Habituation of response to one stimulus exhibits stimulus generalization to other stimuli.	X	X	--	O	--	--	--	--	--	O*
8. Presentation of another stimulus results in response recovery. (Dishabituation)	X	X	X	X	X	X	--	X	O	O
9. Repeated application of dishabituating stimulus causes less recovery. (Habituation of dishabituation)	X	X	--	X	--	--	--	--	O	O

X - Yes   O - No   -- - Not known   * - changes from original table by Eisenstein and Peretz (1973).

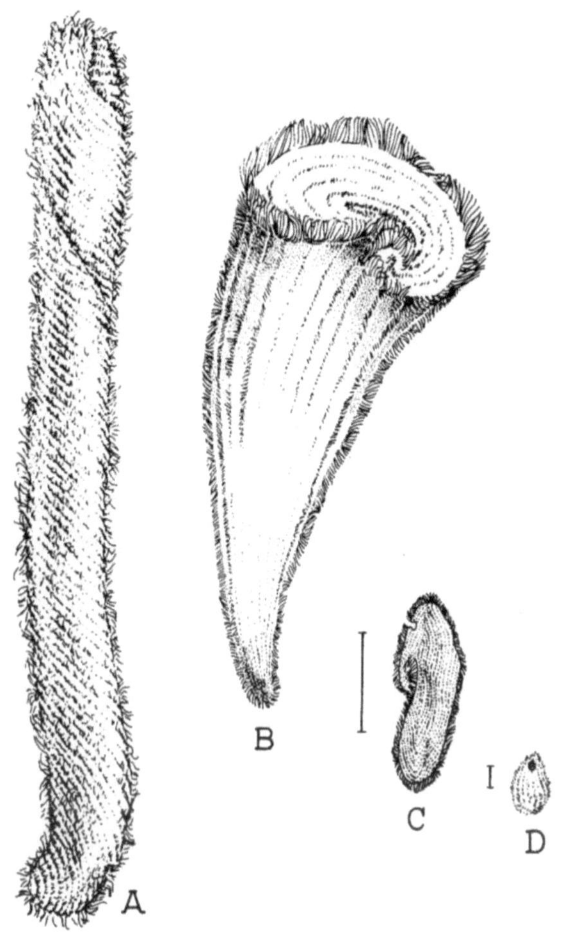

Figure 1. Illustration of some ciliated protozoans. A. <u>Spirostomum</u>
<u>ambiguum</u>, B. <u>Stentor</u> <u>coeruleus</u>, C. <u>Paramecium aurelia</u>, D. <u>Tetrahymena</u>
<u>pyriformis</u>. <u>Spirostomum</u> and <u>Stentor</u>, which can reach lengths of
3 and 2 mm, respectively, are among the largest ciliated protozoans.
The bars next to <u>Paramecium</u> and <u>Tetrahymena</u> indicate what the actual
sizes of these animals would be relative to <u>Spirostomum</u> and <u>Stentor</u>.

a five minute rest period to determine the retention of the habituation. As can be seen, some spontaneous recovery has occurred after
five minutes. Other studies (Kinastowski, 1963, a, b; Applewhite
and Morowitz, 1966, 1967; Thompson <u>et al</u>., 1973) have shown that the
frequency, intensity, and total amount of stimulation affect habituation and its retention in <u>Spirostomum</u> identically to ways found
in neurally mediated habituation (Table 1). Similar results have
been obtained with <u>Stentor</u> (Wood, 1970a).

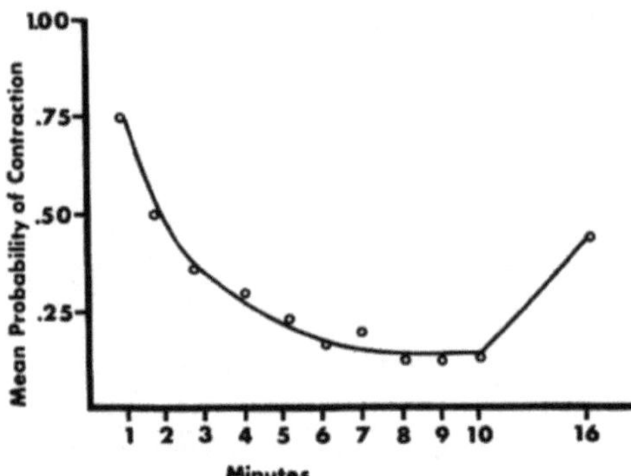

Figure 2. Probability of contraction versus time. The frequency
with which 10 groups of 4 Spirostomum each contracted in response
to 10 minutes of repetitive mechanical stimulation (1 stimulus per
10 seconds) is shown on a per minute basis. After 5 minutes of
rest an additional minute of stimulation was given to test the
retention. The decrease in response probability (habituation) is
significant at the 0.01 level (Wilcoxon matched-pairs test, one-
tailed). (Taken from Osborn et al., 1973).

Habituation in Stentor and Spirostomum occurs at stimulation
rates of about one stimulus per minute or faster, whereas interstimulus intervals used for habituation training involving neural
systems often can be two to three orders of magnitude longer.
Similar differences are found in retention times. Stentor shows
retention for only three to six hours, while Spirostomum shows
retention for less than one hour. This suggests that multicellular
organization may be necessary for long term retention of behavioral
modifications.

Nonetheless, habituation has been conclusively demonstrated
to exist in protozoans, and is therefore not dependent upon the
presence of multicellular organization. However, a pheonomenon
called dishabituation, often observed in neural systems (Table 1),
has not been demonstrated in protozoans (Wood, 1970a; Eisenstein,
unpublished). If a stimulus (often of a higher intensity) of the
same modality, or a stimulus of a different modality, is inserted
out of the previous temporal order of stimulus patterning, the
response level to succeeding stimuli may be affected. If a spontaneous return to or toward the original response level is observed
to the succeeding stimuli given at the original intensity (modality),
then dishabituation or partial dishabituation is said to have
occurred.

Wood (1970a) habituated a group of Stentor for 60 trials at
one stimulus per minute (Figure 3). On the 61st trial the amplitude

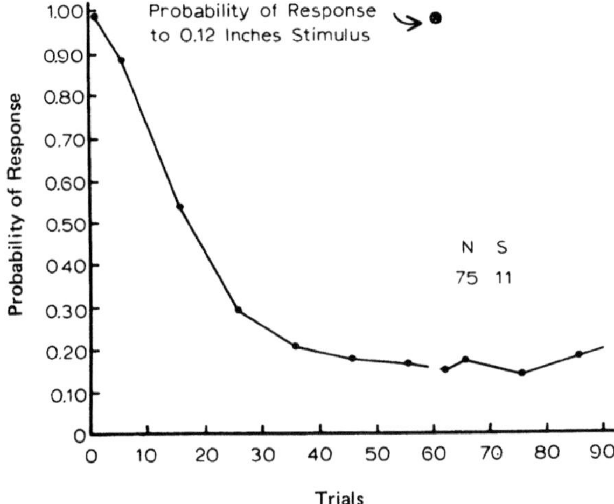

Figure 3. Probability of contraction versus trials. A group of
Stentor (N = 75) were habituated to 60 mechanical stimuli (0.04
inch vertical displacement, 1 stimulus per minute) and on trial 61
given one dishabituating mechanical stimulus (0.12 inch horizontal
displacement). No evidence of dishabituation was found on trials
62-90, which were given at the initial stimulus intensity. (Taken
from Wood, 1970a).

of the mechanical stimulus was increased from 0.04 to 0.12 inches of
test chamber displacement. Response to the dishabituating stimulus
was nearly 100%; however no dishabituation was observed to the
subsequent 0.04 inch stimuli.

Applewhite and Gardner (1971) reported dishabituation in
Spirostomum, but their definition of dishabituation differs from
that usually accepted. That is, they found an increased responsiveness to the dishabituating stimulus, and called this dishabituation. However, the data agree with that of Wood and of Eisenstein's
laboratory in that this increased responsiveness is not found to
subsequent stimuli given at the original intensity. Therefore, as
suggested by Eisenstein and Peretz (1973), perhaps multicellular
organization is necessary for the presence of dishabituation.

Although protozoans are quite responsive to electric shock,
habituation does not occur to electrical in the same time period that
it does to mechanical stimulation in either Stentor or Spirostomum
(Wood, 1970a; Osborn et al., 1973). This provides a convenient
control stimulus with which various properties of habituation can be
investigated. For instance, Osborn et al. (1973) have looked at the
interaction between electrical and mechanical stimulation in Spirostomum. Figure 4 shows the results of giving one minute of electrical

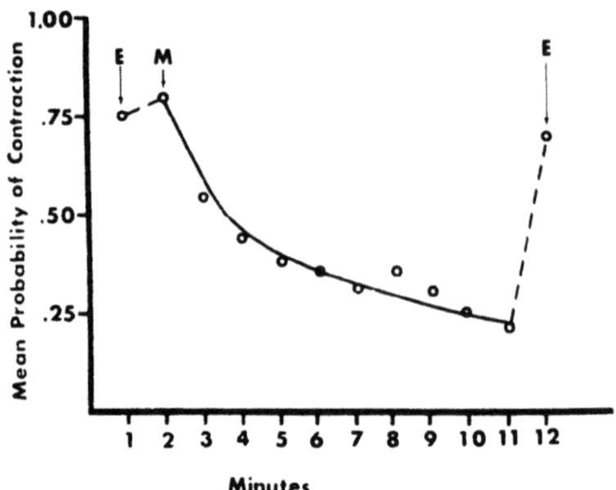

Figure 4. Probability of contraction versus time. After one
minute of stimulation with a 2 msec., 100 volt electrical shock (E),
a group of 20 Spirostomum were given ten minutes (minutes 2-11) of
mechanical stimulation (M). The animals were then given an additional minute of the electrical stimulation. All stimuli were given
at 1 stimulus per 10 seconds. Despite the fact that the animals
habituated to the mechanical stimulation, responsiveness to the
electrical stimulation was not significantly affected. (Taken from
Osborn et al., 1973).

stimulation followed by eight minutes of mechanical stimulation
followed by one minute of electrical stimulation, all at a rate of
one stimulus per ten seconds. As can be seen, habituation occurs
during the eight minutes of mechanical stimulation; but, even
though the animals are habituated to the mechanical stimulus, the
responsiveness to electrical stimulation is not significantly
affected. Figure 5 shows the results of interchanging the modalities
of stimulation. No habituation is seen to either of the two intensities of interposed ten minutes of electrical stimulation, but
responsiveness to the terminal minute of mechanical stimulation is
significantly depressed relative to the initial level in both
cases. However, in neither case is the responsiveness depressed
as sharply as it would have been had only mechanical stimulation
been used. (Note the superimposed mechanical habituation curve.)
Since the group of animals which received the higher intensity of
electrical stimulation elicited more contractions than the group
which received only mechanical stimulation (dashed curve), it is
clear that effector fatigue is not the cause, or at least not the
sole cause, of habituation. Rather, as hypothesized by Osborn
(1971), habituation to mechanical stimulation may occur at the
mechanoreceptor-transducer site. The electrical stimulation may
bypass this receptor-transducer site and activate the contractile
system directly. Wood's (1970b, c) electrophysiological data suggest
a similar situation in Stentor.

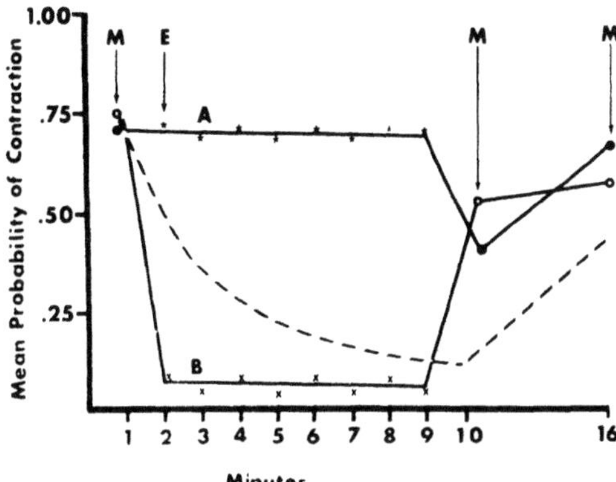

Figure 5. Probability of contraction versus time. After one minute
of stimulation with mechanical stimulation (M), two groups of 40
Spirostomum each were given eight minutes (minutes 2-9) of electrical
stimulation (E). Group A was given 2 msec., 100 volt stimuli, and
Group B, 2 msec., 40 volt stimuli. Both groups then received one
additional minute (minute 10) of mechanical stimulation. One minute
of mechanical stimulation was also given after a 5 minute rest to
test for retention. Although neither group showed a response decrement to the electrical stimulation, responsiveness to mechanical
stimulation was depressed ($p \leq 0.05$, Wilcoxon matched-pairs test,
one-tailed). However the data from Figure 2, dashed line, show that
the response decrement is not as great as it would have been had only
mechanical stimulation been given. (Taken from Osborn et al., 1973).

The fact that, as shown in Figure 5, even low levels of electrical stimulation cause a decrease in responsiveness to mechanical
stimulation suggests that the mechanoreceptor is affected by the
electrical stimulation, either directly or via the effector activity.
Similar results have been obtained in Stentor by Wood (personal
communication).

This interaction between electrical and mechanical stimulation
questions whether or not a protozoan is capable of habituating
independently to differing modalities of stimulation. That is, if
an animal is habituated to one modality of stimulation, will it be
prehabituated to a second stimulus modality to which it normally
habituates? Data from Table 1 show that multicellular systems do
generalize two differing modalities of stimulation and that following habituation to one, responsiveness to the other modality of
stimulation is decreased. Since Stentor habituates to both light
and mechanical stimulation, it provides an excellent system in
which to investigate this problem. Wood (1973) has compared habituation curves obtained from Stentor with mechanical stimulation both
before and after habituation to photic stimulation, as well as light

habituation curves both pre- and post-mechanical habituation. In
neither case was habituation affected by previous stimulation of the
other stimulus modality, thus demonstrating that the habituating
(receptor-transducer) mechanisms for these two modalities of stimulation are independent. This absence of stimulus generalization is
another parameter which may distinguish protozoa from habituating
neural systems (Table 1).

From these behavioral studies it is clear that protozoa are
capable of the learning-related phenomenon of habituation and are
also capable of responding differentially to different modalities
of stimulation. Since the molecular mechanisms regulating these
capabilities are all confined to a single cell, it seems reasonable
to assume that molecular level investigations designed to elucidate
the bases of these behavioral capabilities should be relatively
straight forward. However, although use of a free-living, single-
celled organism for behavioral experiments provides several
experimental simplifications, disadvantages are also present.

Because one cell must carry out the range of functions which
are spread among different cells in multicellular systems, such a
cell is necessarily more complex than a differentiated metazoan
cell. In addition, even though protozoans are individual cells
and as such are free from intercellular connectivities, if large
groups of animals are used for testing, interactions between animals
can still occur. Thompson et al. (1973) have shown that Spirostomum,
habituated four animals at a time, showed no differences from
animals tested individually. However, the relatively crowded conditions which resulted from testing groups of 15 to 20 animals
concurrently in a 6 mm diameter slide well caused inter-animal
interactions to become evident. Animals which were in close physical
proximity would tend to contract in unison, whereas animals in
smaller, less crowded groups would contract independently. Thus,
any study of contractile habituation using large groups of Spirostomum should take this factor into account. This problem is circum-
vented in Stentor, however, which can remain sessile during habituation training.

Chemical interactions are also possible, but control experiments
which (1) tested the responsiveness of naive animals in medium
previously used to habituate other groups of Spirostomum and which
(2) habituated experimental animals in a perfusion slide that per-
mitted a continuous replacement of the medium, both indicate that
chemical (medium) effects are negligible. Also, the components
of these single cells monitor and regulate numerous behaviors (e.g.,
in Spirostomum the animals must concurrently regulate the rate and
direction of ciliary beating, internal contractions of the excretory
contractile vacuole, possible local modifications of the body wall
which result in bending as well as initiation of total contractions

which can shorten the animal to about 40% of its resting length
within 5 msec). Thus, protozoa possess a behavioral complexity
which contrasts sharply with single neurons of multicellular organisms, many of which have the rather discrete function of conducting impulses from one site to another.

In general, the use of protozoans tends to simplify behavioral
observations and to facilitate control of experimental conditions.
However, chemical studies designed to elucidate the molecular
events which control behaviors and behavioral modifications are
complicated by the complex behavioral and physiological organization of these multifaceted cells. For protozoans are indeed also
whole organisms; and as such have more than one ongoing and perhaps
more than one modifying behavior. Thus, before the results of
any biochemical investigations can be meaningfully interpreted,
a complete behavioral framework of events occurring to a given
stimulation pattern must be established. Such investigations
have been undertaken on _Spirostomum_ (Hamilton, 1972; Hamilton _et al._,
1974). Some interesting results have emerged.

For instance, not all of the animals of a large group of
Spirostomum given habituation training actually habituate. Some
of the animals facilitate (increase in contractile responsiveness).
Data from 40 animals, tested individually with mechanical stimula-
tion and given the same stimulus intensity are shown in Figure 6.
Those animals which contracted to four or five of the first five

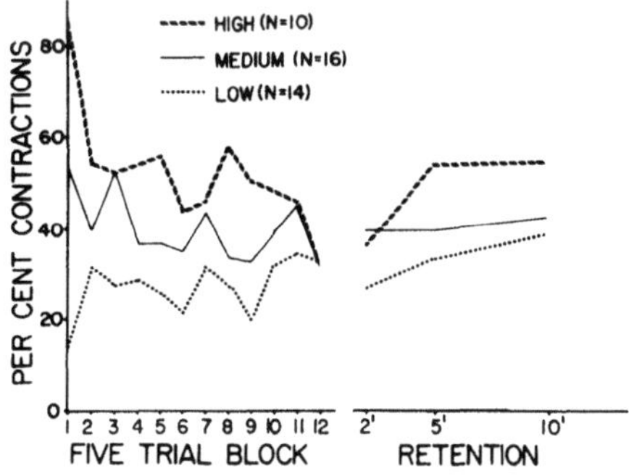

Figure 6. Percent contractions versus trials. _Spirostomum_ given 10
minutes of mechanical stimulation (1 stimulus per 10 seconds) were
classified on the basis of how many times they contracted to the
first 5 stimuli. Those animals contracting 4 or 5 times were designated high responders, those contracting 0 or 1 times were called
low responders. The high responders habituated ($p \leq 0.05$, Daniel's
test for trend, one-tailed) whereas the low responders seemed to
facilitate ($p \leq 0.005$, Wilcoxon matched-pairs test, one-tailed).

stimuli were classified as high responders. Those which contracted
only once or not at all were designated low responders. As can be
seen, the high responders habituate ($p \leq$ 0.05, Daniel's test for
trend, one-tailed) whereas the low responders appear to facilitate
between the first and second five trial block ($p \leq$ 0.005, Wilcoxon
matched-pairs test, one-tailed). Thus, two competing processes,
habituation and facilitation can occur simultaneously in any large
group of _Spirostomum_ given repetitive mechanical stimulation. In
fact, Groves and Thompson (1970) have suggested that in neural
systems this is normally the case. Any molecular level investigations of habituation should therefore take this factor into
account.

Another parameter found subject to change during repetitive
mechanical stimulation is the magnitude of those contractions
which occur. The resting length of _Spirostomum_ was found to remain
constant throughout the entire period of repetitive mechanical
stimulation, but an approximately 10% decrease in the magnitude of
contractions (i.e., an increase in the contracted length) was
observed in those animals which maintained a high frequency of
contracting. The decrease in contraction magnitude occurred over
a 10 to 20 trial interval during which a high percentage of contractions were observed, but the rate of reextension remained
constant. As an animal's contraction probability habituated and
the animal thereby responded less frequently, the contraction
magnitudes returned to the initial level. Thus, another variable
must be considered in interpreting biochemical results from large
groups of animals. That is, the contraction magnitude of high
responders is variable and decreases by about 10% during the early
phases of habituation training, whereas low responders are found
to maintain a constant level of contraction magnitude throughout
an entire stimulation period.

The situation becomes even more complicated when the ciliary,
as well as the contractile system is considered. For, as mentioned
earlier, ciliates like _Spirostomum_ will, besides contracting,
respond to stimulation by reversing the direction of ciliary beat
and swim backwards. Such a response, called an avoidance reaction,
can occur concurrently with or independently of contractions.

It was determined (Hamilton _et al_., 1974) that the probability
with which avoidance reactions were elicited remained relatively
constant during habituation training, but that the amount of time
spent backing up as well as the distances backed up on successive
avoidance reactions decreased. Figure 7 indicates the magnitude,
measured both in terms of the distance backed up and the duration
of the ciliary reversals, of avoidance reactions which were
exhibited in response to the first five or six stimuli during
habituation training. Data are shown for three animals with

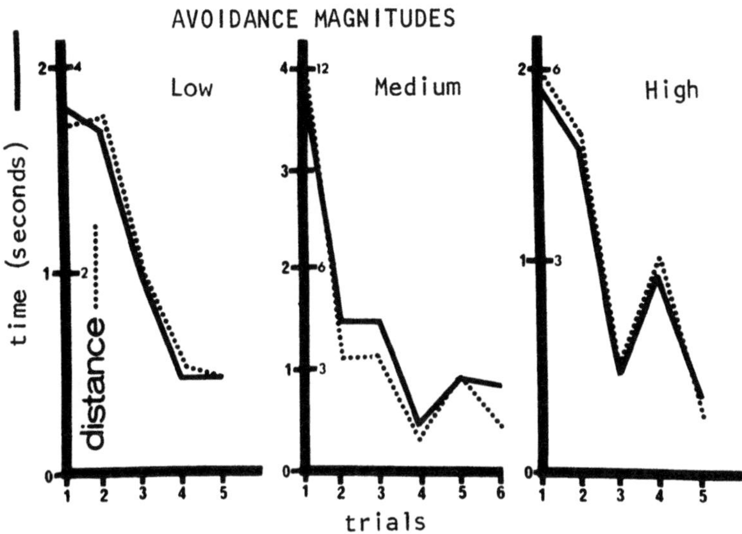

Figure 7. Representative avoidance reaction magnitude changes over
trials. Both the distance backed up and the time spent backing up
on successive avoidance reactions are shown for three animals (one
for each response group) for the first 5-6 trials of mechanical
stimulation. An obvious and rapid decrease in the strength of
avoidance reactions occurs during this time period. Since the time
and distance lines coincide, the rate of backward swimming remains
constant.

differing initial contraction probabilities. Since both the time
and distance lines coincide for each animal, the rate of backward
swimming remained constant. But a response decrement occurred over
only a few trials, much more quickly than contractile habituation,
suggesting that the two processes are independent. Thus, the
interpretation of molecular investigations is not only complicated
by more than one modification within the contractile system, but
also by concurrent changes within the ciliary system. When it is
further considered that response modifications vary among animals
within any large group, it becomes apparent that the "simple system"
approach gained by using protozoans for behavioral experiments does
not necessarily extend to experiments designed to explain the
molecular nature of these behaviors.

Nonetheless, investigations are simplified by the fact that all
relevant events occur in an identified, isolated cell and by the
fact that the environment of this cell can be accurately regulated.
This advantage is not even realized in most of the "simple system"
invertebrate preparations in which habituation has been demonstrated
(Eisenstein and Peretz, 1973). Furthermore, complicated isolation

or preparation procedures are not required to free experimental
cells from other living tissue.

The ability to accurately regulate the animals' environment
has been used by several investigators (Applewhite, 1968b, c;
Gardner and Applewhite, 1970; Thompson et al., 1973) to study
the effect of growth temperature and rapid temperature changes
on habituation. These studies have indicated that initial
contractile responsiveness is temperature independent, but that
habituation is improved by higher temperatures (30°C) whereas
retention is improved by lower temperatures (13°C).

Applewhite et al. (1969b) have attempted to specify the locus
of habituation in Spirostomum. By centrifuging Spirostomum at
2000 g for 13 minutes, the macronuclei, which function in metabolic
control, were forced into one end of the animal. By then cutting
the organism in half with a razor blade, they were able to obtain
pieces of animals that were devoid of all macronuclei and about
1/3 of the cytoplasm. After 10 minutes of rest these Spirostomum
"shells" were subjected to habituation training. Despite the
rather brutal experimental treatment, the "shells" demonstrated
the presence of habituation as well as spontaneous recovery after
the cessation of stimulation. Thus, the cortical layer of these
animals which includes the contractile filaments and a dense layer
of mitochondria, is implicated in the habituation process.

Similarly, Applewhite (1968a) found that if Spirostomum were
cut in half immediately following habituation training to mechanical
stimulation, both halves of the organism retained the habituation
(i.e., showed a decrease in responsiveness) relative to Spirostomum
halves from non-habituated, naive animals. Thus, the locus of
habituation in Spirostomum appears to be generalized throughout most
or all of the cortical layer, rather than localized in one spot.
A recent study by Wood (1972) using a local tactile stimulus has
shown that habituation is also generalized to multiple receptor
surfaces in Stentor.

Other studies, reviewed in detail by Corning and VonBurg (1973),
have investigated both RNA and protein turnover during habituation
training. These studies (Applewhite, 1970; Applewhite and Gardner,
1968, 1970; Applewhite et al., 1969a) indicated that protein synthesis was not necessary for habituation, since an increase in
protein synthesis observed in normal Spirostomum after 50 stimuli
was not observed in Spirostomum "shells" that also showed habituation; but an RNA increase was observed in both normal and "shell"
animals during the early phases of habituation training (20 stimuli).
The increased RNA turnover was no longer evident during the later
stages of habituation training (50 trials) nor during retention.
These RNA turnover data are similar to results discussed by Corning

(1971) for a metazoan, planaria; and thus suggest that RNA may be implicated in _Spirostomum_ habituation.

ASSOCIATIVE CONDITIONING

Attempts to demonstrate classical and avoidance conditioning in protozoans have been many and varied. Corning and VonBurg (1973) have provided an extensive review of these studies, which have been marked by controversy over experimental controls, inability to replicate results, and failure to extend the reported results.

Difficulties in unequivocally demonstrating associative conditioning in protozoans stem from a number of different problems. First of all, the conditioned stimulus (CS) and/or the unconditioned stimulus (US) may affect the environment in a manner that alters an animal's responsiveness independent of any physiological changes within the organism; e.g., electric shock may induce local pH changes and thereby affect responsiveness, elasticity, and/or migratory preferences. Similarly, the animal's physiology may be altered, sensitized, by the stimulation in such a manner that it becomes responsive to the previously neutral CS independent of any CS-US pairing. A final problem, not often considered seriously enough, is that it takes time to learn how to handle protozoans without affecting behavioral experiments. Stated perhaps more clearly, considerable experience with a new system is often necessary before one knows what and when something should or should not be done with experimental animals. For instance, _Spirostomum_ show alterations in contraction probability for several days after subculturing (Thompson _et al._, 1973). Similarly, when transferred into a slide well for behavioral testing, a _Spirostomum's_ contraction probability stabilizes after only five minutes rest, but swimming (ciliary) behaviors are affected for nearly ten minutes. In a small, shallow slide well, the surface of the medium (whether it is flat, convex, or concave) and the magnitude of variation from flatness have significant effects upon contractile responsiveness and other behavioral responses. Also, _Spirostomum_ usually clump in large numbers around wheat seeds that are often added to the culture medium to support the bacterial population upon which _Spirostomum_ feed. Animals taken from these dense clumps differ behaviorally from those found swimming freely in the culture medium (Hamilton, 1972). Many of these handling and culturing phenomena seem obvious, once they are detected, but they are often overlooked and therefore neglected. Unfortunately, most of these "helpful hints" for obtaining consistent results from behavioral testing are never published. Thus, it is difficult to evaluate whether the inability to replicate associative learning results from a seemingly insignificant difference in handling, culturing techniques, chemical

composition of the medium [perhaps resulting from differences in
laboratory stills (e.g., Favero et al., 1971) or local water
sources], etc.; or whether the original findings were merely induced
by some experimental idiosyncrasy of the investigator.

The relatively recent experiments of Bergström (1968a, b, 1969a,
b) are a case in point. In a seemingly well controlled study
Bergström presented light (CS)-electrical shock (US) pairs to
Tetrahymena (Figure 1) which had been grown in the dark at 25°C for
two days. Control groups which received shock only, light only,
or neither shock nor light were also run. After 1, 6, 11, 16, 21,
26, or 31 stimulus presentations the animals were transferred to a
test chamber which was lighted in such a manner that only 30% of
the chamber was illuminated. The results, shown in Figure 8,
indicate that the experimental group avoided light, whereas all
three control groups demonstrated no light avoidance. Retention of

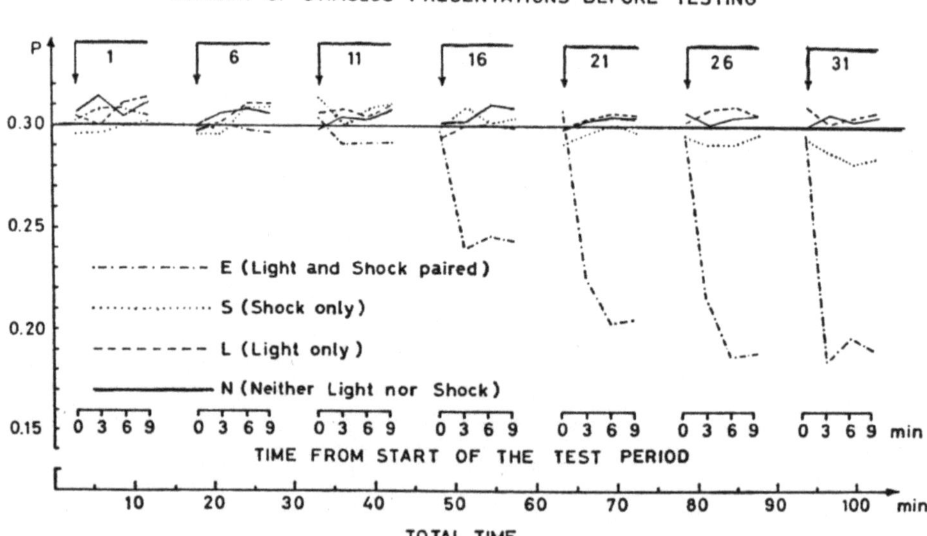

Figure 8. Proportion (P) of animals in the lighted regions of the
test chamber after various numbers of stimuli. Tetrahymena were
given light-shock pairs every three minutes. After 1, 6, 11, 16, 21,
26, or 31 stimulus pairs animals were transferred to a test chamber,
80% of which was lighted. Light avoidance preferences of animals
from each of these groups were then measured for 9 minutes each. An
increased light avoidance was observed relative to control animals
which received shock only, light only or neither light nor shock.
(Taken from Bergström, 1968b).

the light avoidance was dependent upon the number of light-shock
pairings presented, and for 30 pairings the effect persisted for
several hours (Figure 9). His last study (Bergström, 1969b) also
found that a gradual light-dark boundary in the test chamber was more
effective in demonstrating the avoidance conditioning than was an
abrupt boundary. However, Applewhite et al. (1971), who added one
control group to the experiment which received random, unpaired
light and shock, were unable to replicate any of Bergström's findings.
Thus, like the previous half century of work on associative learning
in protozoa, the results of these intriguing experiments are in
controversy, but are indeed suggestive enough to warrant further
investigation.

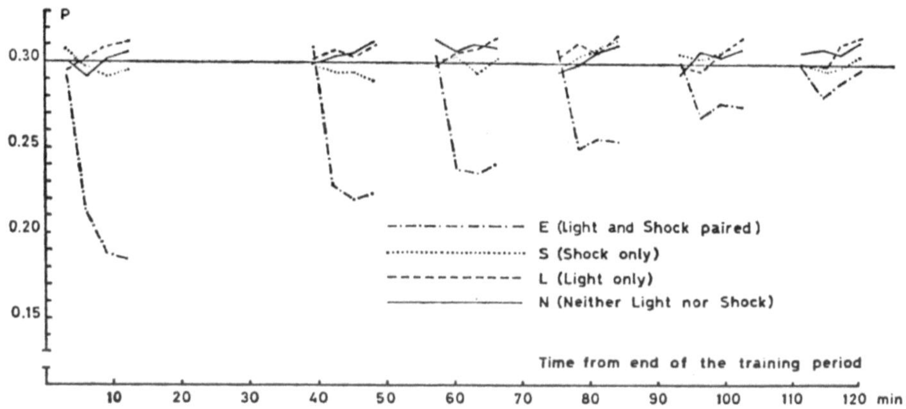

Figure 9. Retention of light avoidance over time. Tetrahymena
given 30 light-shock pairs were transferred to test chambers at
various intervals after training to determine the degree of
light avoidance retained. The experimental group showed increased
light avoidance for about 2 hours relative to control groups that
received shock only, light only, or neither shock nor light.
(Taken from Bergström, 1969a).

ACKNOWLEDGEMENTS

The author expresses thanks to Pauline West for her aid in the
preparation of Figure 1 of this chapter. Research by the author
cited in this chapter was carried out under NIH general research
support grant 5-501-RR-05656-04 to the College of Human Medicine,
Michigan State University; MH grant 1 RO3 MH 18570-01 and NSF grant
GB 23371 to Dr. E.M. Eisenstein. The author was supported by NIH
training grant GM-01422-06. Travel to the Aneural Organism

Symposium at the Winter Conference on Brain Research and the preparation of this chapter were made possible by support the author
received under NIH training grant 5T01 GM 00836-10.

REFERENCES

Applewhite, P. 1968a. Non-local nature of habituation in a rotifer
  and protozoan. Nature, 217, 287-288.
Applewhite, P. 1968b. Temperature and habituation in a protozoan.
  Nature, 219, 91-92.
Applewhite, P. 1968c. Retention of habituation in a protozoan
  improved by low temperature. Nature, 219, 1265-1266.
Applewhite, P. 1970. Protein synthesis during protozoan habitua-
  tion learning. Commun. Behavioral Biol., 5, 67-69.
Applewhite, P. and Gardner, F.T. 1968. RNA changes during pro-
  tozoan habituation. Nature, 220, 1136-1137.
Applewhite, P. and Gardner, F.T. 1970. Protein and RNA synthesis
  during protozoan habituation after loss of macronuclei and
  cytoplasm. Physiol. Behavior, 5, 377-380.
Applewhite P. and Gardner, F.T. 1971. A theory of protozoan
  habituation. Nature: New Biology, 230, 285-287.
Applewhite, P., Gardner, F.T., Foley, D., and Clendenin, M. 1971.
  Failure to condition Tetrahymena. Scand. J. Psychol., 12, 65-67.
Applewhite, P., Gardner, F.T., and Lapan, E. 1969a. Physiology
  of habituation learning in a protozoan. Trans. N. Y. Acad. Sci.,
  31, 842-849.
Applewhite, P., Lapan, E., and Gardner, F.T. 1969b. Protozoan
  habituation learning after loss of macronuclei and cytoplasm.
  Nature, 222, 491-492.
Applewhite, P. and Morowitz, H.J. 1966. The micrometazoa as model
  systems for studying the physiology of memory. Yale J. Biol. Med.,
  39, 90-105.
Applewhite, P. and Morowitz, H.J. 1967. Memory and the microinverte-
  brates. In: Chemistry of Learning, W.C. Corning and S.C. Ratner
  (Eds.), Plenum Press, New York, pp. 329-340.
Bergström, S.R. 1968a. Induced avoidance behavior in the protozoa
  Tetrahymena. Scand. J. Psychol., 9, 215-219.
Bergström, S.R. 1968b. Acquisition of an avoidance reaction to
  light in the protozoa Tetrahymena. Scand. J. Psychol., 9, 220-224.
Bergström, S.R. 1969a. Amount of induced avoidance behavior to
  light in the protozoa Tetrahymena as a function of time after
  training and cell fission. Scand. J. Psychol., 10, 16-20.
Bergström, S.R. 1969b. Avoidance behavior to light in the protozoa
  Tetrahymena. Scand. J. Psychol., 10, 81-88.
Blättner, H. 1926. Beitrage zur Reizphysiologie von Spirostomum
  ambiguum Ehrenberg. Archiv. für Protistenkunde, 53, 253-311.

Corning, W.C. 1971. Recent studies of learning and its biochemical correlates in protozoans and planarians. In: The Biology of Memory, G. Adam (Ed.), Plenum Press, New York, pp. 101-119.

Corning, W.C. and VonBurg, R. 1973. Protozoan Learning. In: Invertebrate Learning, Vol. I. Protozoans through annelids, W.C. Corning, J.A. Dyal, and A.O.D. Willows (Eds.), Plenum Press, New York.

Eisenstein, E.M. and Peretz, B. 1973. Comparative aspects of habituation in invertebrates. In: Habituation: Behavioral Studies and Physiological Substrates, Vol. 2, M.J. Herz and H.V.S. Peeke (Eds.), Academic Press, Inc., pp. 1-34.

Favero, M.S., Carson, L.A., Bond, W.W. and Petersen, N.J. 1971. Pseudomonas aeruginosa: growth in distilled water from hospitals. Science, 173, 836-838.

Gardner, F.T. and Applewhite, P. 1970. Temperature separation of acquisition and retention in protozoan habituation. Physiol. Behavior, 5, 713-714.

Groves, P.M. and Thompson, R.F. 1970. Habituation: A dual-process theory. Psych. Rev., 77, 419-450.

Hamilton, T.C. 1972. A quantitative analysis of behaviors during habituation training of the ciliated protozoan, Spirostomum ambiguum. Ph.D. Thesis, Michigan State University, East Lansing, Michigan.

Hamilton, T.C., Thompson, J.M. and Eisenstein, E.M. 1974. Quantitative analysis of ciliary and contractile responses during habituation training in Spirostomum ambiguum. Behav. Biol., 12, 1-15.

Harris, J.B. 1943. Habituatory response decrement in the intact organism. Psychol. Bull., 40, 385-422.

Jennings, H.S. 1901. Studies on reactions to stimuli in unicellular organisms. IX. On the behavior of fixed Infusoria (Stentor and Vorticella) with special reference to modifiability of protozoan reactions. Amer. J. Physiol., 8, 23-60.

Jennings, H.S. 1906. The Behavior of Lower Organisms. Columbia Univ. Press, New York.

Kinastowski, W. 1963a. Der Einfluss der mechanischen Reize auf die Kontraktilitat von Spirostomum ambiguum. Acta Protozool., 1, 201-222.

Kinastowski, W. 1963b. Das Problem "des Lernes" bei Spirostomum ambiguum. Acta Protozool., 1, 223-236.

Osborn, D. 1971. Contractility in the protozoan Spirostomum ambiguum. Ph.D. Thesis, Michigan State University, East Lansing, Michigan.

Osborn, D., Blair, H.J., Thomas, J. and Eisenstein, E.M. 1973. The effects of vibratory and electrical stimulation on habituation in the ciliated protozoan, Spirostomum ambiguum. Behav. Biol., 8, 655-664.

Thompson, J.M., Hoy, D. and Eisenstein, E.M. 1973. The effects of growth temperature and rapid temperature change on the contractile behavior of Spirostomum ambiguum. In preparation.

Thompson, R.F. and Spencer, W.A. 1966. Habituation: a model
  phenomenon for the study of neuronal substrates of behavior.
  Psych. *Rev*., *73*, 16-43.
Wood, D.C. 1970a. Parametric studies of the response decrement
  produced by mechanical stimuli in the protozoan, *Stentor coeruleus*.
  J. *Neurobiol*., *1*, 345-360.
Wood, D.C. 1970b. Electrophysiological studies of the protozoan,
  Stentor coeruleus. *J*. *Neurobiol*., *1*, 363-377.
Wood, D.C. 1970c. Electrophysiological correlates of the response
  decrement produced by mechanical stimuli in the protozoan, *Stentor
  coeruleus*. *J*. *Neurobiol*., *2*, 1-11.
Wood, D.C. 1972. Generalization of habituation between different
  receptor surfaces of *Stentor*. *Physiol*. *Behav*., *9*, 161-165.
Wood, D.C. 1973. Stimulus-specific habituation in a protozoan.
  Physiol. *Behav*., *11*, 349-354.

PLANT AND ANIMAL BEHAVIOR: AN INTRODUCTORY COMPARISON

Philip B. Applewhite

Associate Professor of Biology
S.U.N.Y. College at Purchase, Purchase, N.Y.
and Visiting Fellow in Biology
Yale University, New Haven, Connecticut

This is intended as a brief introduction to the comparative
aspects of animal and plant behavior with a view as to how plants
as aneural systems may contribute to or expand our understanding of
neurobiology. This is not intended as a comprehensive review of the
subject, as I have taken only selected examples from a vast, though
separate, literature on plants and animals, and have referred to
review papers or books as summary references where applicable.

It would be a mistake to say that plants should be studied
instead of animals if one wants to understand the functioning of
the nervous system. Yet, the parallels are quite intriguing.
I think what emerges from a consideration of plant and animal
behaviors is that they complement each other nicely, and, if nothing
else, demonstrate again just how similar all free-living organisms
can be to one another in comparisons at the molecular level. One
could, of course, be equally struck by the dissimilarities between
plants and animals, making a case that discrepancies from the
expected are more interesting because isolated cases offer unique
perspectives. However, the stronger case seems to come from
similarities since they allow generalizations to be made.

Clearly, if we become basic enough, we can point to plant and
animal similarities such as: they both rely on ATP as an energy
source, are composed by weight mostly of water, have genes composed
of nucleic acids and for another example, utilize enzymes in
metabolic processes. While obvious, these are not trivial similarities as far as life on earth is defined, for they help to define
it here. On some other planet in some other galaxy, these similarities might not exist between this world and that. The fact that

there is a basic similarity here suggests there may also be
similarities in the complex behaviors of plants and animals that,
after all, are built upon these basic principles of life.

Early scientists were not unaware of the behavior of plants.
Fechner (1848) compared <u>Mimosa</u> (Sensitive plant) movements with
animal reflexes. Darwin (1881), too, was particularly interested
in plant movements. Of particular interest here is his discussion
of sleep movements in plants - the closing of leaflets and flowers
to darkness. As he points out this is a very conspicuous behavior
of plants, and references to it occur in the <u>Natural History</u> of Pliny
the Elder, <u>c</u>. 50 A.D. Sleep movements were described for over 85
genera by Darwin; among them, <u>Mimosa</u>, <u>Phaseolus</u> (Bean), <u>Passiflora</u>
(Passion flower) and <u>Hibiscus</u>. Darwin was quite accurate in saying
that with <u>Mimosa</u>, for example, the leaflet closure and dropping of
the petiole (stem) as a response to dark were due to turgor changes
in the motor organs at the base of the leaflets and petiole.
However, he was not willing to make the obvious analogy of plant
sleep movements with animal sleep and was quite specific about this.
Discussions were also presented on geotropism and heliotropism
(now called phototropism).

The Indian biologist Bose (1906, 1913, 1926, 1928) took plants
into the laboratory for detailed investigations of irritability and
was very much taken with the analogy to animals. He looked at the
effect of mechanical stimulation, electric shock, temperature, light
and chemical agents upon plant behavior using <u>Mimosa</u>, <u>Biophytum</u>,
<u>Neptunia</u> and <u>Desmodium</u> as primary objects of study. With regard to
<u>Mimosa's</u> leaflet and petiole movements, he looked at the effects of
stimulus intensity, inter-trial intervals and fatigue on the response
and its recovery. He was, in essence, looking at habituation. In
addition, he did electrophysiology on the surface of plants, recording
from wires attached to stems and leaves, and observed changes
accompanying plant movement. Bose discovered that ether and
chloroform vapor would exert a narcotic effect upon <u>Mimosa</u>,
weakening its rapid closure response to mechanical stimulation.
Wallace (1931, a,b,c) continued this approach with <u>Mimosa</u>. Bose's
investigations over many years.led him to conclude that there was a
"fundamental unity of physiological responses in plant and animal".

Further support to this unity was provided by the carnivorous
plants. Darwin (1875) described observations and experiments on
these plants, concentrating mainly on <u>Drosera</u> (Sundew), <u>Dionaea</u>
(Venus' fly-trap), <u>Utricularia</u> (Bladderwort) and their relatives.
His main interest was in the capture and digestion of the insects
along with the triggering devices that moved the appropriate plant
organs upon stimulation. There are many other early studies of the
carnivorous plants and they are best presented by Lloyd (1942).
For example, a 1910 study (Brown and Sharp) concentrated on the

sensitive trigger hairs of <u>Dionaea</u> to elucidate the sensory aspects
of closure (less than 1 second) of the trap. They found that at
15°C. two stimuli (either touching the same trigger hair twice or
two different hairs once each) were required, within an interval
of 1.5 to 20 seconds, to provoke closure. But, at 40°C. only one
stimulus was usually sufficient. If the interstimulus interval
were extended beyond 20 seconds, more stimuli were needed to close
the trap. When the interval was 3 minutes, 9 stimuli were necessary.
Chemicals may act as the stimulus for movement reaction in the
fungus <u>Dactylella</u> (Couch, 1937). The fungus grows in the form of
rings and when a nematode attempts to swim through one of them,
the ring suddenly swells, trapping the organism. Hyphal bodies
would eventually penetrate the nematode and destroy it. Tactile
and heat stimulation of the fungus seemed to be ruled out as the
triggering stimuli in further experiments, while applications of
lactic acid sometimes produced a swelling response. Unfortunately,
relatively little detailed work utilizing intracellular recording
techniques has been done on the sensory physiology of carnivorous
plants, but diverse studies are being carried out now with these
plants (Jaffe, 1973; Williams & Pickard, 1972a,b).

It is no surprise that in the past, biologists studied the
behavior of plants just as they studied the behavior of animals
for they were all living things. They were sometimes treated
simultaneously as in studying phototropisms (Loeb, 1918), irritability
(Bose's work), or behavior as part of comparative psychology
(Fuller, 1934; Warden, Jenkins and Warner, 1935, 1940). The Warden
<u>et al</u>. reference deserves special mention as it is truly a
comparative behavior study, encompassing all kingdoms of organisms.
Not everyone at the time was willing to say, as Fechner and Bose were,
that in many cases the underlying physiology of plant and animal
responses was identical or that microorganisms had a psychic life
of sorts as Binet (1889) would suggest. Superficially, the
resemblances were there. Animals and plants both showed geo and
phototropisms, "sleep", rhythms, rapid movements, and irritability.
If we go beyond the behavior <u>qua</u> behavior and look at the underlying
physiology and biochemistry, we can better assess these similarities
and develop some new ones from a contemporary perspective.

Not all behaviors are similar; for example, we do not know the
actual biochemical events causing phototropism in plants or phototaxis in animals but they appear different, at least at the levels
at which they have been analyzed. For plants, the plant growth
hormones, the auxins, act on cell elongation (Ball, 1969) causing
differential cell expansion producing a bending to one side.
Phototaxis in animals has been studied best in microorganisms with
as yet no definitive molecular explanation of the phenomenon
(Halldal, 1970). The mechanism may relate to biochemical changes
in surface membranes, brought about by photon absorption, causing

changes in ciliary and flagellar action. The time lag involved in
light response may also suggest some fundamental differences since
animals respond faster than plants, as is also the case in geotropism.
For this latter response, the statocysts in animals seem to be
fundamentally different in their sensing mechanisms than the one
occurring in plants with the displacement of starch grains
(amyloplasts) caused by gravity field (Gordon and Cohen, 1971).
As with phototropisms, auxins exert differential effects on cell
growth after being activated in some way by the amyloplast movements.
One organism that might bridge the gap between the plant and animal
kingdoms in determining just how similar or different their reactions
are to light and gravity is to be found in the Fungi kingdom:
Phycomyces. It apparently does not use auxin transport as part of
the phototropic response (Castle, 1966) nor does it have gravity
receptors similar to plants (Bergman et al., 1969); yet it exhibits
geo and photo tropisms.

     Rapid movements in plants relate quite nicely to animal studies
as both action potentials and actomyosin-like proteins have been
implicated for some species. Rapid plant movements occur in a
variety of species (Sibaoka, 1969) such as some of the carnivorous
plants, Mimosa, some filaments of stamens and the stigma of pistils
and in tendrils of some plants (Jaffe and Galston, 1968). Some of
these rapid movements are preceded by action potentials (Sibaoka,
1969) and include Mimosa leaflet closure, stamen filament movement
in Sparmannia (African hemp) and Berberis (Barberries), pistil
stigma movement in Incarvillea, trap closure in Dionaea, tentacle
movement in Drosera (Williams and Pickard, 1972 a,b) and cucumber
tendril movement (Umrath, 1934). With the exception of the stamen
filaments, the action potentials are also propagated. In addition,
action potentials of longer duration not associated with rapid
movement have been found in the algae Nitella and Chara
(Higinbotham, 1973). What look like action potentials also occur
in shoots and phloem of some plants having no motor activity
(Pickard, 1972; Sinyukhin and Gorchakov, 1966). It is not clear
what purposes these have yet. There is also evidence that
actomyosin-like proteins may be directly involved in plant movements
as they are in muscles in the animal systems. Pisum (Pea) tendrils
(Jaffe and Galston, 1968), Mimosa (Sibaoka, 1969) and Dionaea
(Jaffe, 1973) are the best examples of possible actomyosin involvement, but by no means prove the case. Since plant movement results
from differences in turgor pressure in appropriate cells of the
plant, the actomyosin-like substances may effect the elasticity or
permeability of cells such that water movement occurs.

     The neurotransmitters acetylcholine (Jaffe, 1970; Satter,
Applewhite and Galston, 1972), serotonin and norepinephrine
(Applewhite, 1973) have been found in a wide variety of plants but
their functions, if any, in plant movement has yet to be elucidated;

although, acetylcholine has been implicated in photomorphogenesis
(Yunghans and Jaffe, 1972). Possibly they may function in affecting membrane permeability as they do in other non-nervous tissues
such as red blood cells. Detailed work is currently being carried
out in plant cholinesterases (Fluck and Jaffe, 1974). In the
nervous system, the influx of calcium following action potential
propagation causes the release of neurotransmitters and it has
been proposed that this release may involve actomyosin. It could
cause conformational changes in synaptic vesicle membranes and
pre-synaptic membranes resulting in the direct movement of transmitter from vesicle to synaptic cleft (Berl, Puszkin and Nicklas,
1973). The necessary ingredients for a similar series of events
are present in Mimosa as it has actomyosin-like proteins (Sibaoka,
1969), norepinephrine (Applewhite, 1973) and calcium release
accompanying movement (Toriyama and Jaffe, 1972). A direct interaction of these components has not yet been demonstrated. Further
analogies with neuromuscular systems come from an investigation of
different drugs on the sensitivity of Mimosa to mechanical stimulation (Applewhite, 1972c). The neuromuscular blocking agent curare
and the local anesthetic xylocaine prevent leaflets from closing to
touch and mechanical stimulation. The effect is reversible as
washed leaflets soon close to stimulation. This parallel with
animal systems is most intriguing and may reflect cell membrane
action.

As is well known, animals and plants show endogenous circadian
rhythms and their similarity in responses to light and temperature
has been well documented (Bünning, 1973). Several species of plants
and animals have been shown to generate circadian rhythms in protein
and RNA synthesis and in some cases to have rhythms disrupted by
inhibitors of RNA and protein synthesis (Bünning, 1973; Sweeney,
1969; Brown, Hastings and Palmer, 1970; Menaker, 1971; Applewhite,
Satter and Galston, 1973). As is often pointed out these studies
do not indicate that the clock oscillator is RNA or protein as they
may only be the link between the clock and the behavior it is
controlling. It is not known what the clock is but recent work
suggests that rhythmic changes in membrane permeability may be
important both for animals and plants in producing rhythmic
behavior (Bünning, 1973; Njus et al., 1974; Satter, Applewhite
and Galston, 1974; Sweeney, 1974). For one phase of the rhythm the
membrane state may be such that substances diffuse in one direction
causing a behavior, and during the other phase the membrane state
may be such that these same substances are actively transported
back in the other direction - causing the reversal of the behavior
and resetting the system. Such processes are consistent with the
hypothesis of Bünning (1973) that an energetic phase of half a day
alternates with a non-energetic phase lasting the other half and
both phases comprise the period of the clock. For example, under
constant darkness, the leaflets of Albizzia open for about 12 hours
and close for about 12 hours. The rhythmic opening phase is energy

dependent while the rhythmic closing phase is generally energy
independent (Satter and Galston, 1973). Accompanying these phases
are potassium ion movements in opposite directions; in one direction
by diffusion for closing, in the other by active transport for
opening. Water flow follows the direction of potassium movement
and changes the turgor pressure of appropriate cells with the
resulting opening or closing behavior. Which behavior occurs
depends on relative turgor changes of dorsal to ventral sides of
the pulvinule (the organ controlling leaflet movement). Changes in
membrane permeability are clearly involved here and the question to
be resolved in all these studies is, what causes oscillations in
membrane permeability?

Animal sleep which is certainly rhythmic, shows some interesting
parallels with the energetic-non energetic concept. One theory of
sleep is that among other things it permits the animal to synthesize
substances that are depleted during wakefulness and/or to destroy
toxic substances accumulated during wakefulness. Some evidence has
been found to support this view, as during sleep, compared to wakefulness, there is in the nervous system higher levels of ATP,
protein, acetylcholine and serotonin synthesis (Freeman, 1972;
Hartmann, 1973). Sleep, would be, then, more of an energetic phase,
which it turns out is just the opposite in the "sleep" (closure) of
Albizzia leaflets where the greater protein synthesis occurs during
"wakefulness" (opening) (Applewhite, Satter and Galston, 1973).
Plant "sleep" would effectively slow down energy requirements
(and prevent heat loss) while animal sleep would not, at least not
in the brain.

Another energetic-non energetic combination surfaces in a model
to describe habituation (cessation of contraction) to mechanical
stimulation in the protozoan _Spirostomum_ (Applewhite and Gardner,
1971). It was suggested that the acquisition of habituation
depended upon the diffusion of ions in one direction (away from
myonemes) and the loss of habituation depended upon the active
transport in the other direction. The presence or absence of ions
affects the state of the actomyosin-like proteins present and hence,
the contractility of the organism. This concept was expanded to a
general theory of learning involving ion flow (Applewhite, 1972a)
although not without criticism (Mrosovsky, 1973). Studies of
habituation of leaflet closure in _Mimosa_ (Applewhite, 1973) and
light growth response habituation in _Phycomyces_ (Ortega and Gamow,
1970) indicated that the response characteristics were not unlike
those of animal habituation in many respects. Furthermore, the
biochemical explanation of _Mimosa_ habituation appears quite similar
to that hypothesized for _Spirostomum_, again involving ion flow and
possible involvement of actomyosin-like proteins. This is not
surprising considering their involvement, as discussed above, in
plant movement. It is tempting to speculate that the circadian

rhythms in organisms have something in common physiologically with learning (the acquisition of a response), because both phenomena have been postulated to involve energetic and non-energetic phases of ion movements. Could the entraining of a new rhythm upon an organism be physiologically equivalent to learning? The experiments dealing with learning in plants are assessed elsewhere (Applewhite, In press).

Our knowledge in the areas I have touched upon is moving at a rapid rate. It should not be too long a time from now until we can determine with some accuracy just how biochemically similar animal and plant behavior is. Further research should tell us to what extent the present interface between them is spurious or important, and, therefore, suggestive of common evolutionary processes to deal with the environment in similar ways.

REFERENCES

Applewhite, P.B. 1972 a. The flow of ions in learning and memory.
  J. Theoret. Biol.,36, 419.
Applewhite, P.B. 1972b. Behavioral plasticity in the sensitive
  plant, Mimosa. Behavioral Biol.,7, 47.
Applewhite, P.B. 1972c. Drugs affecting sensitivity to stimuli in
  the plant Mimosa, and the protozoan Spirostomum. Physiol. and
  Behavior, 9, 869.
Applewhite, P.B. 1973. Serotonin and norepinephrine in plant
  tissues. Phytochemistry, 12, 191.
Applewhite, P.B. In press. Learning in bacteria, fungi and plants.
  In Invertebrate Learning, Vol. 3, Ed. by W.C. Corning et al.,
  Plenum Press. New York.
Applewhite, P.B., and Garner, F.T. 1971. Theory of protozoan
  habituation. Nature New Biology,230, 285.
Applewhite, P.B., Satter, R.L. and Galston, A.W. 1973. Protein
  synthesis during endogenous rhythmic leaflet movement in Albizzia.
  J. General Physiol.,62, 707.
Ball, N.G. 1969. Tropic, nastic and tactic responses. In Plant
  Physiology, F.C. Steward, Ed. Vol. VA. Academic Press, New York, 119.
Bergman, K., et al. 1969. Phycomyces. Bacteriological Reviews,33, 99.
Berl, S., Puszkin, S. and Nicklas, W.J. 1973. Actomyosin-like
  protein in brain. Science, 179, 441.
Binet, A. 1889. The Psychic Life of Micro-Organisms. The Open
  Court Publishing Co., Chicago.
Bose, J.C. 1906. Plant Response as a Means of Physiological
  Investigation. Longmans, Green and Co., London.
Bose, J.C. 1913. Researches on the Irritability of Plants.
  Longmans, Green and Co., London.
Bose, J.C. 1926. The Nervous Mechanism of Plants. Longmans, Green
  and Co., London.

Bose, J.C. 1928. The Motor Mechanism of Plants. Longmans, Green
    and Co., London.
Brown, F.A., Hastings, J.W. and Palmer, J.D. 1970. The Biological
    Clock: Two Views. Academic Press, New York.
Brown, W.H. and Sharp, L.W. 1910. The closing response of Dionaea.
    Bot. Gaz.,49, 290.
Bünning, E. 1973. The Physiological Clock. Springer-Verlag, New
    York.
Castle, E.S. 1966. Light responses of Phycomyces. Science 154, 1416.
Couch, J.N. 1937. The formation and operation of the traps in the
    nematode-catching fungus, Dactylella. J. Elisha Mitchell Sci. Soc.,
    53, 301.
Darwin, C. 1875. Insectivorous Plants. John Murray, London.
Darwin, C. 1881. The Power of Movement in Plants. D. Appleton &
    Co. New York.
Fechner, G.T. 1848. Nanna, oder Über das Seelenleben der Pflanzen.
    Voss, Hamburg.
Fluck, R.A. and Jaffe, M.J. 1974. Cholinesterases from plant tissues.
    III Distribution and subcellular localization in Phaseolus aureus.
    Roxb. Plant Physiol. ,53, 752.
Freeman, F.R. 1972. Sleep Research. Charles C. Thomas, Springfield,
    Illinois.
Fuller, H.J. 1934. Plant behavior. J. General Psychol., 11, 379.
Gordon, S.A. and Cohen, M.J. eds. 1971. Gravity and the Organism.
    University of Chicago Press, Chicago.
Halldal, P. 1970. Photobiology of Microorganisms. John Wiley and
    Sons, New York.
Hartmann, E.L. 1973. The Functions of Sleep. Yale University Press,
    New Haven, Conn.
Higinbotham, N. 1973. Electropotentials of plant cells. Ann. Rev.
    Plant Physiol., 24, 25.
Jaffe, M.J. 1970. Evidence for the regulation of phytochrome-
    mediated processes in bean roots by the neurohumor acetylcholine.
    Plant Physiol.,46, 768.
Jaffe, M.J. 1973. The role of ATP in mechanically stimulated rapid
    closure of the Venus's Fly-Trap. Plant Physiol. ,51, 17.
Jaffe, M.J. and Galston, A.W. 1968. The Physiology of tendrils.
    Ann. Rev. Plant Physiol.,19, 417.
Lloyd, F.E. 1942. The Carnivorous Plants. Ronald Press Co., New
    York.
Loeb, J. 1918. Forced Movements, Tropisms, and Animal Conduct.
    J.B. Lippincott Co., Philidelphia.
Menaker, M. Ed. 1971. Biochronometry. National Academy of Sciences,
    Washington.
Mrosovsky, N. 1973. Temperature and learning in poikilotherms.
    J. Theoret. Biol., 39, 659.
Njus, D., Sulzman, F.M. and Hastings, J.W. 1974. Membrane model for
    the circadian clock. Nature, 248, 116.
Ortega, J.K.E. and Gamow, R.I. 1970. Phycomyces: Habituation of
    the Light growth response. Science, 168, 1374.

Pickard, B.G. 1972. Spontaneous electrical activity in shoots of
    Ipomoea, Pisum, and Xanthium. Planta, 102, 91.
Satter, R.L., Applewhite, P.B. and Galston, A.W. 1972. Phytochrome-
    controlled nyctinasty in Albizzia julibrissin. V. Evidence against
    acetycholine participation. Plant Physiol., 50, 523.
Satter, R.L., Applewhite, P.B. and Galston, A.W. 1974. Rhythmic
    potassium flux in Albizzia: Effect of anomophylline, cations and
    inhibitors of respiration and protein synthesis. Plant Physiol.
    In Press.
Satter, R.L. and Galston, A.W. 1973. Leaf movements: rosetta stone
    of plant behavior? Bioscience, 23, 407.
Sibaoka, T. 1969. Physiology of rapid movements in higher plants.
    Ann. Rev. Plant Physiol., 20, 165.
Sinyukhin, A.M. and Gorchakov, V.V. 1966. Action potentials of
    higher plants not possessing motor activity. Biofizika,11, 966.
Sweeney, B.M. 1969. Rhythmic Phenomena in Plants. Academic Press,
    New York.
Sweeney, B.M. 1974. The potassium content of Gonyaulax polyedra
    and phase changes in the circadian rhythm of stimulated biolumi-
    nescence by short exposures to ethanol and valinomycin. Plant
    Physiol., 53, 337.
Toriyama, H. and Jaffe, M.J. 1972. Migration of calcium and its
    role in the regulation of seismonasty in the motor cell of Mimosa
    pudica L. Plant Physiol., 49, 72.
Umrath, K. 1934. Über die elektrischen erscheinungen bei thigmischer
    reizung der ranken von cucumis melo. Planta,23, 47.
Wallace, R.H. 1931a. Studies on the sensitivity of Mimosa pudica.
    I. The effects of certain animal anesthetics upon sleep movements.
    Amer. J. Botany,18, 102.
Wallace, R.H. 1931b. Studies on the sensitivity of Mimosa pudica.
    II. The Effect of animal anesthetics and certain other compounds
    upon seismonic sensitivity. Amer. J. Botany,18, 215.
Wallace, R.H. 1931c. Studies on the sensitivity of Mimosa pudica.
    III. The effect of temperature, humidity and certain other factors
    upon seismonic sensitivity. Amer. J. Botany,18, 288.
Warden, C.J., Jenkins, T.N., and Warner, L.H. 1935, 1940. Vol. 1, 2.
    Comparative Psychology. Ronald Press Co., New York.
Williams, S.E. and Pickard, B.G. 1972a. Receptor potentials and
    action potentials in Drosera tentacles. Planta, 103, 193.
Williams, S.E. 1972b. Properties of action potentials in Drosera
    tentacles. Planta,103, 222.
Yunghans, H. and Jaffe, M.J. 1972. Rapid respiratory changes due to
    red light or acetylcholine during the early events of phytochrome-
    mediated photomorphogenesis. Plant Physiol., 49, 1.

CONTRIBUTORS

Philip B. Applewhite
Associate Professor of Biology
S.U.N.Y. College at Purchase
Purchase, N.Y.
and Visiting Fellow in Biology
Yale University
New Haven, Connecticut

Victor Kai-Hwa Chen
Department of Biology
Case Western Reserve University
Cleveland, Ohio

Bodo Diehn
Department of Chemistry
The University of Toledo
Toledo, Ohio

E.M. Eisenstein
Department of Biophysics
Michigan State University
East Lansing, Michigan

Miles Epstein
Department of Zoology
University of Minnesota
Minneapolis, Minnesota

Earl M. Ettienne
Department of Anatomy
Harvard Medical School
Boston, Massachusetts

Thomas C. Hamilton
Central Intelligence Agency
Washington, D.C.

Ching Kung
Department of Biological Sciences
University of California
Santa Barbara, California

David C. Wood
Department of Psychology
University of Pittsburgh
Pittsburgh, Pennsylvania